The Innovators

The Innovators

The Engineering Pioneers Who Made America Modern

David P. Billington

John Wiley & Sons, Inc.
New York ■ Chichester ■ Brisbane ■ Toronto ■ Singapore

To Phyllis and to David

Copyright © 1996 by David P. Billington
Published by John Wiley & Sons, Inc.

Library of Congress Cataloging-in-Publication Data
Billington, David P.
      The innovators : the engineering pioneers who made America modern
  / David P. Billington
          p.   cm.
      Includes bibliographical references and index.
      ISBN 0-471-14026-0 (alk. paper)
      1. Engineering—United States—History.   2. Technological
  innovations—United States—History.   3. Engineers—United States.
  I. Title.
  TA15.B53   1996
  620′.00973—dc20   95-43653

Printed in the United States of America

10 9 8 7 6 5 4 3 2 1

Contents

Preface   *vii*

Part I Iron, Steam, and Early Industry, 1776–1855   1

Chapter 1   Modern Engineering and the Transformation of America   3
Chapter 2   Watt, Telford, and the British Beginnings   21
Chapter 3   Fulton's Steamboat and the Mississippi   41
Chapter 4   Lowell and the American Industrial Revolution   67
Chapter 5   Francis and the Industrial Power Network   79

Part II Crossing the Continent, 1830–1883   95

Chapter 6   The Stephensons, Thomson, and the Eastern Railroads   97
Chapter 7   Henry, Morse, and the Telegraph   119
Chapter 8   St. Louis versus Chicago and the Continental Railroads 139
Chapter 9   Carnegie and the Climax of Steel   159
Chapter 10 Edison and the Network for Light   179
Chapter 11 The Centennial Revolutions, 1876–1883   199

Notes and References   *221*
Index   *237*

Preface

This book is an engineering history of the United States that differs in two ways from most histories of American technology.

First, the book is written from an engineering perspective rather than from the standpoint of social history, and it is also selective in its emphasis on those engineering innovations that were basic to the industrialization of the United States: large-scale structures, prime movers, wide-area networks, and large-scale processes.

Second, the book also focuses on the roles of key figures. The engineering pioneers of U.S. history could have done little without a society that supplied them with capital, trained personnel, and abundant materials and that strongly encouraged new enterprise. Although it is not intended to be a general history of technology in the United States, this book will explain how engineering ideas drew on the unique conditions that existed in nineteenth-century America. At the center of the story, though, are a handful of talented individuals. Where many textbook accounts would lead students to believe that the great industries of the United States were the work of financiers and thieves, this book rescues from historical oblivion the engineers who actually built major industries and in many cases ran them ethically, such as the railroad builder J. Edgar Thomson.

The book treats U.S. engineering history as an interplay of three perspectives: what great engineers actually did, the political and economic conditions within which they worked, and the influence that these designers and their works had on the nation. This three-sided view implies technical discussion, historical context, and cultural impact. We shall discover that the essence of engineering lies not just in natural science, as is usually thought, but also in social science and the humanities. We will explore the scientific basis of engineering through elementary formulas, its social context through issues of politics and economics, and its cultural significance in terms of its impact on the imagination and experience of artists and critics.

The narrative text briefly describes the history of major engineering events, arranged by topic in roughly chronological order. Included are formulas that clarify these events, but the more detailed discussions of these expressions of scientific ideas are placed in pagelong "sidebars" meant for those readers interested in seeing how the engineers' calculations served as one basis for their designs. The formulas in the text express far more than scientific ideas; they are engineering formulations that also include the social issues of regulated loads, visually striking forms, economy of motion, acceptable risks, the bureaucracy of centralized power supply, environmental issues, the production of wealth, and the private support of culture.

The origins of this book go far back to my undergraduate years in the late 1940s, when I was required to take a course given by our dean, Kenneth Condit, who called it "Industrial Development." He had also invented a program, called "Basic Engineering," in which we took courses in all main branches of engineering. Condit's ideas stuck in my mind so that four decades later, when asked by our associate dean, Ahmet Cakmak, to develop a new freshman course, I began the work that has led to this book. That course now has four main parts: Connecting the Continent, 1776–1883; The Rise of the Great American Industries, 1876–1939; Regional Restructuring, 1921–1964; and Information and Infrastructure, 1946–1996. This book represents the first part of the course, and I plan to prepare a volume for each of the remaining three parts.

Acknowledgments

My teaching from 1958 on has been inspired and sustained by Norman Sollenberger, my primary academic mentor, as well as by the late Joseph Elgin, our dean from 1954 to 1972. I am also indebted to my close Princeton colleague Robert Mark and to John Abel, now at Cornell University. Then, in 1983, support from the Alfred P. Sloan Foundation in its program "The New Liberal Arts" allowed me to begin work on the new course with the firm backing of the foundation's president, the late Albert Rees, and his program officer, Samuel Goldberg. Thanks to their continuing support, I could work with stimulating Princeton colleagues, Mark, Michael Mahoney, and John Mulvey as well as with John Truxal and Marian Visich, Jr. from the State University of Stony Brook. They taught me a great deal as did a group of professors of natural science from liberal arts colleagues: William Case at Grinnell College, Alfonso Albano at Bryn Mawr College, J. Nicholas Burnett at Davidson College, and Newton Copp and Andrew Zanella at the Claremont colleges. That group also included Edwin T. Layton, Jr. from the University of Minnesota who brilliantly guided and instructed us in the history of technology. Robert Prigo of Middlebury College added to our group and crucial help has come from historians Carl Condit, Arthur Donovan, Eugene Ferguson, Brooke Hindle, Thomas Hughes, Ronald Kline, the late John Kouwenhoven, the late Melvin Kranzberg, Ronald Paxton, Robert Post, Walter Vincenti, and Robert Vogel. I also thank Merritt Roe Smith, a longtime friend and advisor as well as Atle Gjelsvik who deepened my understanding of boat design.

Crucial to the development of the course, and hence this book, have been my teaching colleagues at Princeton: Bradley Dickinson, John Gillham,

Richard Golden, Michael Littman, as well as Peter Bogucki, Hal McCulloch, Roy Jackson, Paul Prucnal, and Fred Dryer. Our present dean, James Wei, has continuously and enthusiastically supported this effort. Of great importance has been my six-year scholarly collaboration with politics professor Jameson Doig. Together we taught an American Studies seminar linked to the course and together we have been studying the relationship between engineering and politics in the United States. The National Science Foundation also provided crucial support.

I have had gifted graduate students working both as teaching assistants and research assistants during the evolution of this book, beginning with Scott Hunter and including Christopher Peck, Ron Wakefield, Rosemary Secoda, John Matteo, Roger Haight, and Karen Mielich. Susan Lyons and Nicholas Edwards have not only served in the course but also have worked directly on this present book acting as effective collaborators on its final form.

For all their help, I also want to thank my editors at Wiley, Charity Robey and Emily Loose, as well as Jennifer Ballentine and the staff of Professional Book Center, who put the book together.

My longtime friend and collaborator, J. Wayman Williams, has done research, made slides, desktop-published monographs, organized exhibitions, and for this book developed the sidebars as well as provided me with a challenging critique of the text. My secretary, Kathy Posnett, has committed many versions of the text to hard copy while keeping things running well in the office.

Finally, my family deserves the highest praise for their help and sustenance. My brother, Jim, a distinguished historian, has given me both an understanding of historical scholarship and a model of personal and professional integrity. My son, Philip, who is a writer, drafted several chapters in early versions of the book and my daughter, Sarah, who is an engineer, read critically some chapters. The greatest help of all came from my son David, an historian, who carefully read the entire text, suggested many changes including reorganization of major parts, gave me a critical assessment of the leading ideas, and did the index. His keen editorial eye and his deep sense of history shaped the final manuscript.

My wife, Phyllis, has given me the constant help that makes everything possible. While I read, calculate, or write, her romantic piano music fills the house and reminds us both that such music itself arose in parallel with the industrial revolution about some of whose themes this book attempts to tell.

David P. Billington
Princeton University

Figure Credits

Fig. 2.7: Photo by J. Wayman Williams. **Figs. 2.8 and 2.9:** Photo by Elizabeth Billington Fox.

Fig. 3.2: Fulton, R., *Treatise on the Improvement of Canal Navigation,* London, 1796. **Fig. 3.3:** Morgan, John S., *Robert Fulton*, New York: Mason/Charter, 1977, following p. 118. (Original image in the Collections of the American Society of Mechanical Engineers.) Reproduced with permission of the ASME. **Fig. 3.4a:** Fulton's notebook. **Fig. 3.4b:** Philip, Cynthia Owen, *Robert Fulton: A Biography*, New York: Franklin Watts, 1985, following p. 180. **Fig. 3.5:** Durant, John & Alice, *Pictorial History of American Ships,* New York: A. S. Barnes and Company, 1953, p. 173. Photo courtesy of the Cincinnati Historical Society. **Fig. 3.7:** Donovan, Frank, *River Boats of America,* New York: Thomas Y. Cromwell Co., 1966, p. 50. **Fig. 3.8:** Meltzer, Milton, *Mark Twain Himself*, New York: Bonanza Books, 1960, p. 34. Reproduced with permission of the author.

Figs. 4.1 and 4.2: Greenslet, Ferris, *The Lowells and Their Seven Worlds*, Boston: Houghton Mifflin Company, 1946. From the Collections of Mrs. J. H. Ropes. Photos courtesy of the Massachusetts Historical Society. **Figs. 4.3 and 4.4:** Photos by David P. Billington. **Figs. 4.5, 4.6, 5.2:** Malone, Patrick M., *Canals and Industry: Engineering in Lowell, 1821—1880*, Lowell, Mass.: Lowell Museum, 1983, pp. 4, 8, 10. Maps originally drawn by Mark M. Howland, 1975. Historic American Engineering Record, National Park Service.

Figs. 5.4 and 5.5: Francis, James B., *Lowell Hydraulic Experiments,* New York: D. Van Nostrand, 1871.

Fig. 6.1: Photo by J. Wayman Williams. **Fig. 6.3:** Warren, J. G. H., *A Century of Locomotive Building by Robert Stephenson & Co.: 1823–1923*, Andrew Reid & Co. Ltd., Newcastle-Upon-Tyne, 1923, p. 176. **Fig. 6.4:** Newhall, Beaumont, *The History of Photography*, New York: Museum of Modern Art, 1964, p. 72. Photo reproduced with permission from the Collections of the George Eastman House, Rochester, New York. **Fig. 6.6:** Ward, James A., *J. Edgar Thomson: Master of the Pennsylvania*, Westport, Conn.: Greenwood Press, 1980, p. 92. Photo originally from the Division of Archives and Manuscripts, Pennsylvania Historical and Museum Commission, Harrisburg, Pennsylvania. Reproduced courtesy of The State Archives of Pennsylvania. **Figs. 6.7 and 6.8:** Ward, James A., *J. Edgar Thomson: Master of the Pennsylvania*, Greenwood Publishing Group, Inc., Westport, CT, p. 31, 103. Copyright © 1980.) Reprinted with permission of Greenwood Publishing Group, Inc.

Fig. 7.2: Billington, David P. and Alfonso M. Albano, "The Steamboat and the Telegraph," *Episodes in American Invention*, Monograph Series of the New Liberal Arts Program, Princeton, N.J., 1990, p. 57.) Reproduction courtesy of J. W. Williams Associates. **Figs. 7.3, 7.4, 7.5:** Princeton University collection. Photograph by J. Wayman Williams.

Figs. 8.2 and 8.3: Woodward, C. M. *A History of the St. Louis Bridge*, St. Louis: G. I. Jones and Company, 1881, title page, plate XIX. **Fig. 8.4:** Stover, John F., *American Railroads*, Chicago: University of Chicago Press, 1961, p. 85. Reprinted with permission.

Fig. 9.3: McHugh, Jeanne, *Alexander Holley and the Makers of Steel,* Baltimore, Md.: The Johns Hopkins University Press, 1980, p. 225. Originally from the Collections of Mrs. Ashbel Wall. Reproduced with permission.

Fig. 11.2: Roebling, John A., *Long and Short Span Railway Bridges*, New York: D. Van Nostrand, 1869, title page.)

Figures 2.2, 3.1, 3.6, 6.2, 6.5, 7.1, 7.6, 7.7, 8.1, 9.1, 9.2, 10.1, 10.2, 11.3: Courtesy of the Library of Congress. **Figs. 5.1:** Smithsonian Institution Photo No. 91- 6376; **5.3:** Smithsonian Insitution Photo No. 91-6375. Reprinted with permission; **2.3 and 11.1:** Courtesy of the Smithsonian Institution, The Engineering Collection.

I

Iron, Steam, and Early Industry, 1776–1855

1 Modern Engineering and the Transformation of America

Formulas in History

This book tells the story of some major events during the first century of the United States seen through the lens of modern engineering. My central thesis is that engineering has transformed not only the material life of our nation but also its politics and its culture. Although such a thesis is hardly new, I have chosen to include, as a natural part of the story, the engineering description of the transforming events, the central facts of which must be expressed in quantitative terms through some fundamental formulas of the engineer.

My goal is to use these formulas to clarify and deepen readers' understanding of history, not to give readers an engineer's facility with calculations. Because history, even the history of technology, is normally written without such formulas, I need to explain why they are central here.

Formulas are, literally, symbolic images of physical relationships. The formulas presented in this book are rigorously correct but radically simplified versions of what engineers today regularly use to design modern works. These relationships do not involve any mathematics beyond high school algebra, geometry, and trigonometry and usually employ very little of those subjects, either. These formulas are intended to be accessible.

These relationships provide a means of understanding the origins of main events in our history. When Leonard Gale showed Samuel Morse how to get enough magnetism to his receiver, the telegraph became a reality. Gale knew what to do because he knew the relationships among magnetic flux, current, and turns of wire around the soft iron core of the magnet. He knew the formula, and this knowledge, combined with Morse's skill as a designer and organizer, eventually led to a major event in the history of the United States: the innovation of telegraphic communication.[1]

The formula for magnet flux or for stresses in the walls of boilers gives us a new means of understanding and evaluating events in the modern world, not only of the nineteenth century but of the late twentieth century as well, where electric power generation and containment vessel safety are political issues critical to modern society.

These formulas carry more meaning than scientific relationships: They also convey social meaning by their implications of danger and of economy. It is easy to see in them the way in which reductions in materials or increases in pressures lead to competitive economies and to greater risks of failure. The dialogues between the experts and the public, between the engineer and the politician, between the designer and the journalist, all require a common language free from jargon but enlivened by the clarity of formula.

Finally, it is crucial for the public, politicians, and journalists to know that formulas do not solve problems. Formulas suggest designs, stimulate insights, and define limits, but they never provide ways to the best solutions, as so many technologically illiterate writers on engineering suppose. Formulas do not define a "one best way" or an optimum. Formulas represent a discipline, not a design; they can be used to avoid disasters but they can never ensure full safety or essential elegance.

Transformations

In the mid-1820s a young immigrant from England, Thomas Cole, painted an American landscape as wilderness, a painting without sign of people, where the terror of the untracked scene expressed raw nature. Two decades later, Cole, by then a leading American painter, created a landscape entitled *The Pic-Nic*. It is the same nature as before except there are people quietly enjoying a summer outing in the woods. There is no wilderness, no terror; nature had been domesticated.[2] What caused the change in vision? What transformed Cole's landscape and by the late 1840s the landscape of the northeastern United States?

The steamboat and the railroad brought people into the forest and transported Cole himself up the Hudson Valley, where he had moved from New York City. With those new vehicles went also an expanding iron industry and the new electrical telegraphy. Engineering was transforming the United States and with it the physical landscape, the political process, and the artistic sensibility. This book tells the story of that transformation through the central symbols of engineering: formulas.

Formulas are not merely the means for mathematical manipulations. They are the relationships essential to a modern industrial society. They carry scientific meaning, of course, but they also express major social ideas that cannot be exhibited more clearly otherwise. These ideas and these meanings have gradually reconstructed U.S. life over the past two centuries. This book carries the story of that change up to the 1880s, by which time the United States had emerged as the world's leading industrial nation.

This emergence depended first upon a series of major engineering events: the steamboat, the textile factory town, the continental railroad, electric telegraph, the iron and steel industry, the steel bridge, and the incandescent light. But each new event caused immediate political problems unforeseen in the late eighteenth century; engineering innovations made essential new political structures. They also gave rise to a new aesthetic. These interactions between engineering and society form the context within which the relationships of transformation operate. I commence this study of engineering in the modern world with a more detailed exploration of these relationships.

Transformations and Engineering Works

The three modes of transformation—nature, politics, and art—represent the three basic studies in the liberal arts—natural science, social science, and the humanities. Modern engineering, to be properly understood, must be studied in the context of these liberal arts. If the context is the liberal arts, which can be a definition of culture, then what is engineering?

Rather than postulate some abstract and general definition, I propose to follow a more traditional line. People expend great energy, for example, trying to define art. Agreement rarely results. Rather, people successfully teach art history without precise definitions by teaching the works upon whose enduring value the profession agrees. Although there will always be disagreement at the edges, any course in painting will be defective that leaves out the works of Michelangelo, Leonardo da Vinci, and Rembrandt up to Picasso, Klee, and Mondrian.

Similarly, art history can be taught by period: classical, medieval, Renaissance, and so on. In any case it depends not upon scientific-type definitions of theory and applications but rather upon examples and ideas. In addition to individual artists and schools of art, there are also categories of art: painting, sculpture, and architecture as the classical visual arts, for example. In the same way we can identify engineering works through four primary categories: structures, machines, networks, and processes. Furthermore, we can observe that modern engineering in the United States divides itself into periods, the earliest of which corresponds to the development of the prototypical modern material, industrialized iron and steel.

As steel became the building material of the modern world and its offspring, the railroads, became the greatest business of its time, there arose the U.S. industries of oil, electricity, and automobiles. This second period, characterized by central power and private mobility at high speed, dominated the first half of the twentieth century and led naturally into a third period during which the nation turned to major regional restructuring of the landscape through new political instruments such as the Port of New York Authority and the Tennessee Valley Authority. Finally, following World War II, the nation entered the period of high technology that culminated in the microchip and the computer. At the same time, the United States became aware of the deterioration of its built infrastructure of power, water, rails, and roads.

Along with categories and periods comes a set of major events personified by individual pioneers whose innovations have set them apart. Therefore, in tracing the first century of early modern engineering in this book, I focus on the four categories and the people whose works best typify the innovations that transformed the nation.

Raw nature cannot support civilization without four types of transformation: structures for public works; machines for private enterprise; networks for the supply of water, light, power, and signals; and processes for the conversion of natural resources into useful materials. Each type of engineering work has its own character, yet none can exist alone; they are interdependent. But for an introduction each is described separately, and in later chapters each is illustrated by major specific events that changed our society.

Transformation by Structures

Modern structures begin with the use of iron for large-scale works, the first of which was the 1779 Iron Bridge, a cast-iron arch in England. The key formula for understanding this arch bridge is:

$$H = \frac{qL^2}{8d}$$

where H is the horizontal force (in pounds) needed at each of the two side supports to keep the arch from spreading, q is the load on the bridge (in pounds per foot of bridge deck length), L is the horizontal span of the arch (in feet), and d is the vertical rise of the arch (in feet).[*] The figure explains these symbols, which are the same (but with the curve inverted) for the cable suspension bridge.

In the suspension bridge the traffic loads are taken by the horizontal deck to the vertical connecting lines or suspenders that, in turn, pull down on the cables, which carry all the loads by tension (pulling apart) to the tower tops. There the cables push the towers down vertically and pull the towers inward horizontally by the same force, H, given in the formula. To prevent the towers from leaning, the cables go over the towers and are anchored near the deck, usually on the shore. These anchor cables often support the side spans through vertical suspenders as well.

In a cable, the horizontal force H also represents the total cable force at midspan, where the cable itself is horizontal. The larger the force, the more steel the cable requires, so that the equation expresses not only the size of a force but also the amount of material needed for safe design. Moreover, the size of the force H depends directly on the ratio of main span to cable sag, or L/d. Finally, the bridge weight including traffic loads, expressed by qL, directly influences the size of H and hence cable material. The maximum cable tension actually occurs at the tower, but the difference is not great and does not change the following argument.

The scientific fact of the bridge, expressed by the formula, is that vertical loads (qL) are converted into a horizontal force (H) by means of the ratio of span to sag (L/d). That represents a primary discipline of engineering design; that fact is indisputable, and no modern bridge can be built without it. However, the formula gives no hint as to how to proportion the structure, which means the formula cannot serve as a basis for design. We must find another factor, and that comes from the social perspective, which includes cost, benefit, and politics.

The first social issue—qL, the load on the bridge—contains two parts: the traffic load and the weight of the bridge span itself. In the modern world, a public agency sets the traffic load, and that act is political. Will the bridge carry the heaviest trucks or even armored tanks, and how many lanes will it have? Such decisions represent choices bounded by unrestricted use and limited budgets. Large bridges require large public funds for construction, which in a free, democratic society must be justified publicly.

---

[*] In this text I employ the units regularly used in British and U.S. engineering throughout the period covered.

How Structures Work

Arch Bridge

H = horizontal thrust in the arch or cable at midspan (pounds)
q = load (pounds per foot)
L = span (feet)
d = rise of the arch or sag of the cable at midspan (feet)
V = vertical force (pounds)

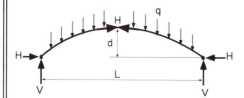

$$H = \frac{qL^2}{8d}$$

$$V = \frac{qL}{2}$$

Cable Suspension Bridge

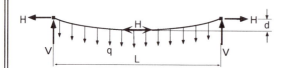

$$H = \frac{qL^2}{8d}$$

$$V = \frac{qL}{2}$$

A second social fact, arising from the ratio L/d, is the cost of towers and cable. Where that ratio is small, then the towers must be high (large d means tall towers) but the cables may be small (large d means small H). The first simple formula tells us these facts, and the size of H controls the amount of steel in the cable through a second formula

$$A = \frac{H}{f}$$

where A is the cross-sectional area (in square inches) of the steel in the cables and f is the allowable stress in the steel (in pounds per square inch of cross-sectional area). The formula is scientific because it expresses a fact of physics; it is independent of any social factors. Having H from the first formula, then for any value of f there will be only one possible solution A.

But the formula also expresses a social fact because someone must choose f. The term *allowable stress* is not a scientific but rather a social statement. Scientifically, a steel wire of a given size, say, $\frac{1}{4}$ inch in diameter (about $A = 0.05$ square inches), will break when pulled by a certain force H, say, 10,000 pounds. That is a scientific fact. The breaking stress f_B would be about 200,000 pounds per square inch, but no one wants a bridge designed such that the real traffic load plus bridge dead weight stresses the cables to the breaking point. Therefore, someone must decide how high the stress can go and thus increase A in order to reduce f. The social consequence of increasing A is greater cost (more steel in the cables), but there is the social benefit of a greater safety factor (the ratio f_B/f). Such factors are set by codes that have legal force and result in the end from choices made by the general public. In the simplest possible terms, the public can get for the same cost three bridges instead of two bridges if it is willing to reduce the safety factor for each of the three by 33 percent (see figure). In a poor country struggling to industrialize, such a decision might be wise, and indeed many U.S. railroad bridges of the nineteenth century were built with low safety factors in order to quickly construct long lines to widely separated settlements.

Finally, the bridge design results not only from scientific facts and social factors, but also from symbolic ideas. These ideas come from individual designers whose image of the bridge transcends all formulas. The best designers see their creations as works of art so that they have aesthetic, ethical, and spiritual meaning. Yet at the same time designers must work within the laws of nature and the patterns of society. Even these symbolic ideas are visible in our simple formulas, although they cannot be fully caught by any formula or indeed by any verbal formulation. The ratio L/d expresses the primary bridge proportion. Contrast the Golden Gate

How a Cable Works

Cable is composed of wires

Cable

H
Horizontal
tension in
pounds

Suspenders

A
Area of
all wires

$$f = \frac{H}{A} \text{ or } A = \frac{H}{f}$$

A = area (square inches)

H = tension (pounds)

f = allowable wire stress (pounds/square inch)

f_B = wire breaking strength (pounds/square inch)

Where A (area) = .05 square inches for each wire and H_B = 10,000 pounds breaking strength per wire

$$f_b = \frac{H_B}{A} = \frac{10,000}{.05} = 200,000 \text{ psi (pounds per square inch) breaking strength}$$

$$\text{Safety factor} = \frac{\text{breaking strength}}{\text{allowable stress}} = \frac{f_B}{f}$$

When one chooses a safety factor of 2: $\dfrac{f_B}{f} = 2$

Thus, allowable stress: $f = \dfrac{f_B}{2} = \dfrac{200,000}{2} = 100,000 \text{ psi}$

If one decides to reduce the safety factor to 1.33, the allowable stress would be:

$$f = \frac{f_B}{1.33} = \frac{200,000}{1.33} = 150,000 \text{ psi}$$

Engineering Choices Are Economic and Political

Two bridges   or   Three bridges

The calculated horizontal tension H in a new bridge is 30,000 pounds per cable.

Safety factor = 2.0 requires 6 wires

Using a safety factor of 2.0, allowable stress:

$$f = \frac{f_B}{2} = \frac{200,000}{2} = 100,000 \text{ psi}$$

Each wire has area $A = 0.05$ square inches, so the allowable force would be $0.05 \times 100,000 = 5,000$ pounds. Thus each cable with 30,000 pounds capacity would require 6 wires.

Safety factor = 1.33 requires 4 wires

If one decides to reduce the safety factor to 1.33, the allowable stress would be:

$$f = \frac{f_B}{1.33} = \frac{200,000}{1.33} = 150,000 \text{ psi}$$

Each wire has area $A = 0.05$ square inches, so the allowable stress would be $0.05 \times 150,000 = 7,500$ pounds. Thus each cable with 30,000 pounds capacity would require only 4 wires.

By reducing the safety factor from 2.0 to 1.33, only two-thirds the amount of material is needed and it might be possible to build a third bridge.

(Large suspension bridges require thousands of wires per cable. For simplicity only a few wires are used in this example.)

Bridge, with a ratio of 8.9, to the George Washington Bridge, having a ratio of 10.8. The West Coast bridge emphasizes high towers whereas the East Coast one has flatter cables. These decisions were not arrived at by scientific or even social analysis but rather by aesthetic desires of the designers. The ratio L/d defines the primary bridge form.

Othmar Ammann, the designer of the George Washington Bridge, put this aesthetic preference strongly with his description of the design for which he "has always been an admirer of the early English suspension bridges with their general simple appearance, their flat catenary, light, graceful, suspended structure, and their plain massive and, therefore, monumental towers."[3]

Transformation by Machines

Modern machines began with the 1775 parliamentary extension of James Watt's steam engine patent, which made economically possible the formation of a private enterprise, the Boulton and Watt Company of Birmingham, England.[4] The formula describing this engine is

$$Hp = \frac{PLAN}{33,000}$$

where Hp is horsepower, defined by Watt to be 33,000 foot-pounds per minute, which he estimated to be the work (in foot-pounds) that one horse could do in 1 minute. He said that it could lift 330 pounds 100 feet high in 1 minute (or 660 pounds 50 feet high in 1 minute, etc.). The power of Watt's engine was created in a cylinder the internal dimensions of which are a length L (in feet) and a cross-sectional area A (in square inches) by an atmospheric pressure P (in pounds per square inch) acting against a piston having the same cross-sectional area A as the cylinder. The figure explains these symbols. In Watt's engine, the separate condenser (see the figure) causes the steam to condense and thereby creates a partial vacuum in the cylinder. This power then comes from a pressure P exerted on the top of the piston and equal to the difference between atmospheric pressure and the reduced pressure (partial vacuum) in the cylinder. In steam engines for later use in steamboats and locomotives, the pressure P came directly from high steam pressure and separate condensers were dispensed with. For fixed power plants, such condensers are still used today.

In the reciprocating steam engine the steam force PA (in pounds) is converted into power by means of the piston motion LN (in feet per minute). The reciprocating motion was a natural one to use for pumping water out of mines. As with the bridge formulas, this power formula expresses a scientific fact and cannot be denied. For a given force PA, a piston motion LN will always produce the same power.[5] This formula underlies all reciprocating machine performance but it says nothing about design. For that we need to look at social factors that, unlike those for bridges, have less to do with public funds and political decisions and more to do with private enterprise and business economics.

We can identify two basic uses for prime movers, or machines that produce power: fixed power plants and mobile power plants. In the former, the machine itself is solidly built into a foundation even though its parts (the piston) move. In the latter, the machine is built into a vehicle

Horsepower from the Cylinder in an Atmospheric Steam Engine

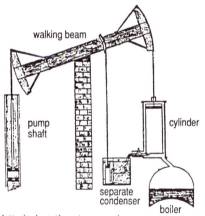

walking beam

pump shaft

cylinder

separate condenser

boiler

P

partial vacuum

L

D

Watt single acting steam engine with separate condenser that creates a partial vacuum

$$A = \frac{\pi D^2}{4}$$

$$Hp = \frac{PLAN}{33,000}$$

Hp = horsepower (33,000 foot-pounds per minute)

P = pressure (pounds per square inch)
    equals atmospheric pressure minus partial vacuum

L = piston stroke (feet)

A = piston head area (square inches)

N = strokes (number per minute)

that itself must move; there the machine produces vehicle movement. The fixed machine delivers power often for industrial consumption, whereas the free machine gives mobility often for individual people. Again the formula tells us some basic social facts about machines.

For fixed power plants, size and weight are not crucial because they sit on cheap concrete and indeed must be prevented from undue vibration. Therefore, they can be large (big L and A) and do not have to have such large internal pressures (P) or operate at such great speeds (N). Fixed power, up until the end of the nineteenth century, meant large engines working at relatively low pressures and speeds. By contrast, first the steamboat and then the locomotive required greater speed and lighter construction, hence small engines (small L and A) with higher pressures and faster actions (greater P and N).

Greater pressures and faster speeds were to machines what greater allowable stresses were to bridges: economy versus safety. The steam had to be contained in boilers and these began to explode, in part because of the increase in steam pressure. Because the motion of a steamboat or locomotive depended upon horsepower, the greater the power, the greater the speed of the boat or the train, and that translated directly into economy. In the competitive market economy of the nineteenth-century United States, faster travel meant higher profits for the transportation companies, whether they were carrying passengers or cargo. Just as structures characterize the politics of public works, so machines typify the economics of private enterprise.

But the machine transformed also the ideas of artists and the general public. Fast steam power replaced laborious human power, and the speed produced by steam gave the average person a new mobility freed from human exertion and opening up seemingly unlimited travel. As Thomas Cole intimated, the new machines changed the public's image of the country and began that westward march that persisted to the end of the twentieth century.

Transformation by Networks

The metal bridge and the steam locomotive typify structures and machines of the nineteenth century and more broadly the two complementary objects of engineering that have arisen since the industrial revolution. Each type of object, however, demands other major engineering works. The iron and steel of the bridge and railroad required new metallurgical processes such as the Bessemer converter to be produced economically, and the rail line itself formed a transportation network throughout vast reaches of territory. That network badly needed traffic control in the 1840s and a new engineering force appeared just then to provide the long-distance signaling essential for safety and scheduling. These two new types of works—networks for transportation and signaling, and processes for the making of iron and steel—are the systems of engineering without which the objects such as structures and machines cannot function.

The prototypical network at midcentury was the telegraph, the first major application of electricity, and it was followed shortly after the centennial by the network for electric lighting. We can characterize these networks by the first simple equation, Ohm's law,

$$V = IR$$

where V is the electromotive force (in volts), I is the current (in amps), and R is the resistance (in ohms). This formula expresses the performance of a circuit, the simplest form of electrical network, in which a generator produces a certain "force," or voltage, and the resistance around the circuit allows a resulting current. Where the resistance is too low (by a short circuit), then the resulting current will become too high, and that high current can lead to a fire or to an electrocution unless the circuit is broken automatically by a safety device (circuit breaker or fuse).

This relationship is again a scientific fact: For 110 volts (the U.S. standard) a circuit resistance of 110 ohms will always result in a current of 1 amp, just as a circuit resistance of 1.1 ohms will invariably give a current of 100 amps. Again the formula says nothing about proper circuit design. For that we need to think about social factors. In a private house 100 amps in a single circuit could be deadly, whereas in some industrial uses it makes good economical sense.

As fixed power plants got larger and hence more powerful, it became economical to try building more extended networks with power delivered to sites at farther distances from the central plant. That desire raised a new problem, expressed through a second major formula

$$P = VI = I^2R$$

where P is the electrical power (in watts) and where, through Ohm's law, power can be written as the product of current squared and resistance. Known as Joule's law, this formula tells us that the power provided to a circuit will be expended in proportion to the various resistances multiplied by the current squared.

This formula provides the basis for lighting where lamps of about 100 watts give strong reading light. But to get that power to a lamp

Networks for Electric Lighting

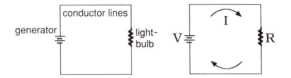

Ohm's law (1)

$$V = I \quad R$$

V	I	R
voltage at the generator	current flowing in the circuit	resistance of the lamp
volts	**amps**	**ohms**

Joule's law (2)

$$P = V \quad I$$

P	V	I
power from the generator	voltage at the generator	current flowing in the circuit
watts	**volts**	**amps**

Substitute (1) into (2)

$$I^2 \quad R = P$$

I^2	R	P
current squared	resistance of the lamp	power converted to light

Note: Formulas for this simplified network neglect the power losses in the lines.

requires electric lines that in themselves consume power due to the resistance of wires. For example, if the generator provides 110 volts and the lamp has a resistance of 110 ohms, but the wires from generator to lamp are so long that there is a total resistance of 990 ohms, then only 0.1 amp will flow and the lamp will have a power only $(0.1)^2 \times 110 = 1.1$ watts, which is not enough to provide useful reading light.

The character of a network is to allow uninterrupted flow. If a wire is cut, current ceases and the network fails. If a bridge collapses, the local railway network ceases to function. Traffic jams and blackouts are the images of broken networks.

The continuity of a network goes together with its flow. In a network the uninterrupted flow must also be an uncorrupted flow. The signal or vehicle that starts its journey within the network must end that journey without distortion, diminution, or destruction. The telegraph signal must be clear, the rail car must be intact, and the fresh water must remain safely potable from reservoir to faucet.

The network operates against isolation. It requires a centralization of power, control, and storage; it demands large-scale engineering construction. Only industrialized societies can afford such systems and only such societies can maintain them.

The U.S. network of rails and telegraphs connected a continent as much as it pushed westward a frontier. Railroads, after all, did not go from east to west after the Civil War, but rather from both coasts inward to meet inland. Always the goal was to connect all parts of the continental territory to make access easier, quicker, and more reliable. That goal symbolizes the network: continuity and transmission.

Transformation by Processes

By contrast to the network, the process represents a system within which things change, being transmuted rather than transmitted. That which enters the system gets refined, broken apart, or assembled with other things. Iron ore, crude petroleum, or automobile parts, each in themselves useless to society, are converted into the products upon which an industrial economy depends.

As an example, we take two simplified expressions for the production of iron in a blast furnace

$$2C + O_2 \rightarrow 2CO$$

where the carbon (C), in the form of charcoal, coal, or coke, combines with oxygen (O_2) to form carbon monoxide (CO). This highly reactive gas (CO) then combines with the iron ore, such as hematite (Fe_2O_3),

$$Fe_2O_3 + 3CO \rightarrow 2Fe + 3CO_2$$

to produce pure iron (Fe) and the waste gas, carbon dioxide (CO_2). Many more reactions take place in the blast furnace but these simple ones give the essence of the process.

Blast Furnace

The Process of Smelting Iron Ore

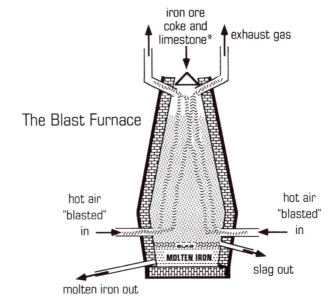

$$2C \; + \; O_2 \; \rightarrow \quad 2CO$$
coke   air blast   carbon monoxide

$$Fe_2O_3 \; + \quad 3CO \quad \rightarrow \; 2Fe \; + \quad 3CO_2$$
iron ore   coke and air   iron   exhaust gas

*Limestone is essential for removing impurities from the iron ore in the blast furnace, but for simplicity it is not shown here in these elementary formulas.

The scientific fact of the second equation is that one molecule of hematite will react with three molecules of carbon monoxide to form two molecules of iron and three molecules of carbon dioxide. This is elementary chemistry as applied to the science of metallurgy. But beyond that scientific fact lie immense social consequences. In simplest form the two products represent the fundamental environmental issue of modern society: an invaluable material (iron) and a potentially dangerous byproduct (carbon dioxide). Industrial society is literally built on ferrous metal while at the same time threatened by gaseous pollution.

Iron ore represents natural resources that must be transformed into useful materials before a modern society can exist. Steel and gasoline made possible the transportation networks, the vehicles, and the cities of the modern world. They also created immense wealth both for a few individuals such as Carnegie and Rockefeller and for the U.S. population as a whole. Yet as the formula makes clear, the benefits of wealth bring with them potentials for degradation, and the balancing of those two facts is the challenge of politics. Private profit and public environment represent another form of the primary domestic issue in U.S. politics. Without widespread wealth, there can be no free, democratic politics, and without that clean public environment there can be no widespread health and enterprise.

The Interconnections of Engineering

These four main ideas in engineering—structures, machines, networks, and processes—are intricately bound together in a modern industrial society. None can exist alone and no industrialized nation can survive if even one is missing. Close study of any one idea will quickly lead to each of the other three.

The imperfect but suggestive metaphor of a human body makes clear how these four ideas coexist in society. We cannot stand without the structure of our skeleton. Bones and joints carry our weight without visible deformations. D'arcy Thompson illustrated this structural metaphor in his suggestive classic *On Growth and Form*.[6]

A skeleton is, however, the image of death unless animated by the heart, a pumping machine without which life stops. But the structure and machine do not make us human. We need also those complex systems that transmit signals and transmute air, water, and food. We can, after all, relatively easily replace parts of the skeleton and even, though it is much more difficult, replace a heart. But the network of electrical signals within and from the brain is far more delicate and elusive. Finally, we cannot survive without the processor of nourishment centered in the stomach that refines food. It is obvious that each major object or system is essential to a healthy life, and so it is in the life of a modern society.

A highway bridge makes no sense without the network of roads leading to it, and the structure itself is made from the steel and concrete that are products of modern chemical processes. Moreover, almost no highway bridge would be built today if there were no motor vehicles. The market for cars, therefore, depends upon the network of roads and the structure of bridges, just as their production depends upon the manufacturing process of an assembly line and ultimately the processes of steel making, among other materials.

The network of electrical power centers in the structures of a power plant, generated by turbo machinery and most often fueled by a process of chemical energy conversion. Similarly, an oil refinery is a series of structures containing and transmitting fluids and gases throughout a piping network and all run by an electronic network centered in computers.

These late-twentieth-century computers are the ultimate in engineering complexity. They permeate society as nothing has ever done. What are they? Surely they are machines, so designated by the giant company IBM. Yet they are also electric networks whose function is to process signals. These almost mysterious inventions are, however, not part of the nineteenth-century story recorded in the following chapters. What computers do in the late twentieth century are, nevertheless, functions that have stimulated invention since the late eighteenth century when modern society began its independent, industrial life in the United States.

2 Watt, Telford, and the British Beginnings

Design: Art and Invention

Two Americans, Robert Fulton and Francis Cabot Lowell, during the early period of independence, went to Britain and brought back technological ideas that were to be central to the development of the early U.S. wealth that made true independence possible. What they found abroad can be characterized by two Scotsmen, James Watt and Thomas Telford, who were the most central figures in transforming Britain from a disconnected and agricultural economy to the greatest industrial state in the world.

I begin therefore with a brief description of these two major figures in the history of modern technology, James Watt (1736–1819) and Thomas Telford (1757–1834). Watt, the engineer of machines, was the greatest engineering inventor of the industrial revolution (roughly from 1775 to 1825), to which "no one has made a greater individual contribution."[1] Telford, the engineer of structures, was the greatest engineering artist of that same period; his bridges, nearly all of which still serve today, are the first works of modern engineering to show a personal aesthetic vision.[2] Moreover, Telford wrote the first modern treatise on engineering as art and consciously thought in artistic terms.[3]

By reviewing some major works of these two designers, we can see why British engineering was so central a starting point for U.S. development. But neither designer could have flourished without the uniquely favorable conditions of the society in which they worked. It is no coincidence that they both worked in Great Britain, nor was it mere chance that they both were Scots. The context in which they worked was the industrial

or British revolution, the hallmark of which was the new material, industrialized iron.

Iron and the Industrial Revolution

The ideas of Watt and Telford came from a century-long improvement in iron making that began in 1709 when Abraham Darby I first
succeeded in the smelting of iron with a mineral fuel, coal. He had moved
in 1708 to Coalbrookdale on the Severn River near Shrewsbury, where
there were abundant supplies of wood, coal, and iron ore nearby. In addition, a small stream falling down the steep valley side supplied water
power. The use of coke (from coal) instead of charcoal (from wood) produced a more fluid iron "that could run freely into smaller channels of the
moulds: The founders could now obtain castings of lighter and more delicate design than had hitherto been possible."[4]

Almost at the same time (1712), Thomas Newcomen (1663–
1729) designed a successful steam engine in Dartmouth for the purpose of
pumping water from mines. While these innovations were being tried out,
a series of inventions (between 1733 and 1779) gradually mechanized the
textile industry in Britain. By 1835, Britain produced more than 60 percent of all cotton goods in the world.[5]

It was in this context that James Watt conceived of the separate
condenser as a basic improvement on the Newcomen engine. He took out
his patent in 1769. Shortly afterward (1774) John Wilkinson, iron founder,
got a patent for a new method of boring cannon. By this method, he was
soon able to bore cylinders for James Watt, and thus overcome one of
Watt's major difficulties in producing a workable steam engine.[6]

Watt and Telford were both Scots, and at just this time Scotland
was the intellectual center of Great Britain. On March 6, 1776, another
Scot, Adam Smith (1723–1790), published *The Wealth of Nations*, the single most significant book on economics up to then.

Three years later, Abraham Darby III, grandson of the pioneer
iron worker, completed the Iron Bridge near Coalbrookdale. The form had
been designed by Thomas Pritchard, architect, and the completed work
was the first large-scale metal structure ever built. Its cast iron arches
spanned 100 feet over the Severn River.

Thus the context was set for Watt and Telford. Steam engines
were in common use before Watt, and iron was produced well before Telford. Moreover, Britain's economy was growing fast and its insularity
helped create a political stability conducive to new developments.

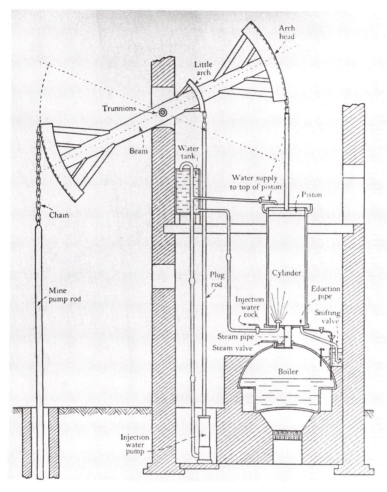

Figure 2.1 The Newcomen engine of 1712

James Watt: The Designer as Inventor

Born near Glasgow in 1736, Watt began his career as a mathematical instrument maker to the College of Glasgow in 1757. Six years later, Dr. John Anderson, professor of natural philosophy (natural science), gave Watt a model of a Newcomen steam engine to repair.[7] While fixing it, he studied carefully how it worked and began to make up his own experiments aimed at improving the engine's efficiency, specifically the amount of steam required to get a certain mechanical action. Watt set about to un-

Figure 2.2 James Watt (1736–1819)

derstand why the Newcomen engine required so much steam (and hence so much fuel), and he soon realized that the problem lay in the cylinder.

The Newcomen engine, primarily used as a pump for draining mines, consisted of three main parts: the boiler in which the heat from the burning fuel converted water to steam; the cylinder, into which the steam entered below the raised piston and was condensed by a jet of cold water; and the beam, a rocking arm supported at midlength, which was pulled down on one side when the condensed steam left a vacuum below the piston. In the Newcomen engine the top of the piston was exposed to the atmosphere and the power (downward) stroke resulted from atmospheric pressure, hence the name *atmospheric engine*, by which the 1712 invention of Thomas Newcomen was known. As the beam rocked, it pulled up the mine pump rod and with it the mine water.

The power of the Newcomen engine results from the *pressure (P)* on the surface area of the piston head (see the previous diagram of the

Newcomen engine). For example, in the best Newcomen engine of Watt's time (designed by John Smeaton in 1772 for the coal mine at Long Benton), the cylinder pressure was about 7.5 pounds per square inch (psi) on the upper side of the piston.[8] Because the cylinder below it was only a partial vacuum, the engine thus developed a pressure of only about half of the total atmospheric pressure of 14.7 psi.

This pressure acted on a *surface area (A)* equal to the area of the 52-inch diameter circular piston head, which for Smeaton's engine was 2,124 inches squared.

The *Force* on the top surface of the piston head is the pressure times the area, which gives here 7.5 psi $\times$ 2,124 in^2 = 15,930 pounds (the square inches cancel) pushing downward on the piston.

This force moves the piston down a total distance L, which we call the *power stroke* of the engine. In the Newcomen engine of 1772 that distance $L = 7$ feet.

This engine produced 15,930 pounds times 7 feet, or 111,500 foot-pounds (ft-lbs) of *work* with one power stroke. In other words, the engine could lift 15,930 pounds of water 7 feet with one full movement of the piston within the cylinder. A horse could lift water out of a mine but with nowhere near as much power.

Power is the amount of work done in a given time, and we find that power by determining how quickly the piston moves, measured by the *number of power strokes per minute (N)*. Our 1772 Newcomen engine could operate at $N = 12$ power strokes per minute, and hence the total power was the work carried out in 1 minute: 111,500 ft-lbs per power stroke times 12 power strokes per minute (min) equals 1,338,000 ft-lbs/min.

James Watt defined power in terms of the capacity of one horse, which he assumed could perform 33,000 ft-lbs of work in 1 minute. Therefore, one *horsepower (Hp)* equals 33,000 ft-lbs/min and the Newcomen engine of 1772 could perform 1,338,000/33,000 = *40.55 Hp*. We can now write the full equation for horsepower:

$$\text{Hp} = \frac{7.5 \times 7 \times 2,124 \times 12}{33,000} = 40.55 \text{ Hp}$$

The units of the numerator must result in ft-lbs/min so P is in pounds per square inches, A is in square inches and L is in feet per power stroke, and N is the power strokes per minute.

The major problem Watt saw with this engine was the rapid swing of temperature within the cylinder. When the steam entered, the cylinder needed to be at 212 degrees Fahrenheit to avoid heat loss, but to condense the steam, the cylinder needed to be about 60 degrees Fahrenheit. Therefore, Watt reasoned, much energy was lost in reheating the cylinder walls for each power stroke. Between late 1763 and May 1765, Watt continued his experiments and his reflections on steam until one Sunday, while he was strolling on the Glasgow Green, the idea of the separate condenser suddenly came to him.

Watt's idea was to add to the engine a fourth major component: a vessel immersed in cold water and into which the hot steam could be fed from the cylinder. This vessel was the condenser, still today a major component of steam-generation power plants, and it permitted the cylinder to remain hot continuously. Ultimately, this reduction in heat loss led to an engine that for the same fuel consumption could produce more than three times as much mechanical work as the best Newcomen engine (Smeaton's design of 1772) and more than nine times as much as the original one (measured in 1718).[9]

The story of how Watt's engine moved from an inspired Sabbath insight to a solid commercial success is one of the most famous in the history of technology. In brief, Watt struggled with his idea off and on for nine years before moving from Glasgow to Birmingham in 1774, where he entered into partnership with the industrialist Matthew Boulton (1728–1809). The combination of Boulton's entrepreneurial flair and Watt's inventive genius allowed the firm of Boulton and Watt to dominate the market for steam engines throughout Britain and abroad for a quarter of a century. The two major factors that made this period (1775–1800) such a success for the firm were the 25-year patent extension granted by Parliament in 1775 and the special laboratory created by Boulton that permitted Watt to continue his inventing.

Watt had taken out a patent for the separate condenser in 1769 that would have run out in 14 years. Boulton realized, when Watt moved to Birmingham in 1774, that he could never recoup his capital investment before the patent monopoly ended in 1783, and therefore he immediately began to lobby Parliament in London to get a law enacted extending the patent. His efforts succeeded and in May 1775, almost exactly 10 years after Watt's original insight, they were granted a 25-year patent extension.

Commercial success required more than a monopoly. It needed the continual technical attention of Watt to make his idea work efficiently at full scale. By relieving Watt of financial worries and providing him with an industrial organization, Boulton in effect created a pioneering research and development operation of the kind that would be so typical of twentieth-century industry.[10] This environment not only allowed Watt to keep inventing (such as the double-acting engine and the rotative engine in 1782,

Figure 2.3 Matthew Boulton (1728–1809)

the parallel-motion idea and the steam carriage of 1784, and the smoke-consuming furnace of 1785), but also put pressure on him to develop working models. Boulton continually pushed the notoriously dilatory Watt into developing practical machines. A primary example was Boulton's insistence on the development, over Watt's objection, of the rotative engine. This engine converted the reciprocating, pumping action of Watt's early steam engine into rotary motion, which could be used to drive textile mills and factories of all kinds. Watt preferred to stay with the reciprocating pump whereas Boulton saw the much wider market for rotating machines to power industry.[11]

Watt was prototypical of the designer as inventor, whose central drive was to improve existing works by making them more efficient.

Figure 2.4 The Watt single-acting engine as presented to
Parliament in 1775

Watt's major inventions involved motion and the desire to control it and create it with as little energy consumption as possible. Furthermore, his inventions centered on works to be mass produced. An improvement, in effect, led directly to a standard.[12]

Design, therefore, especially of machines, involved invention as a major factor but, as the case of Boulton and Watt illustrates, it also involved innovation; that is, the development necessary to turn new insights and ideas into commercially successful objects and systems. Watt, of course, thought about the entire steam engine as a working whole, but he had to spend large amounts of time on the details because each part had to fit accurately and move harmoniously with all others. Watt had to orchestrate the instruments to work in concert.

Watt's biographer refers to him as craftsman and engineer, but it would be more accurate to call him scientist and engineer.[13] Without Boul-

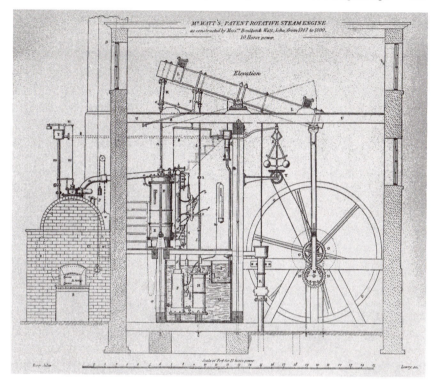

Figure 2.5 Watt steam engine (rotary type)

ton's constant pressure to turn the laboratory ideas into commercially viable machines, Watt's engines would not have revolutionized industry. Indeed, it is not too far off to state that Boulton made a mechanical engineer out of Watt, the would-be scientific researcher.

Thomas Telford: The Designer as Artist

Thomas Telford was born in 1757 at Glendinning near Langhold in southern Scotland, where he was raised by his widowed mother in great poverty. At 13, he was apprenticed to a local stonemason. He left his local valley in 1780, spending two years in Edinburgh as a mason, and then moved to London in 1782, where he planned to become an architect. His design talent attracted enough attention to secure him in 1787 the position of Surveyor of Public Works for the County of Salop, where, mainly as an architect, he designed and supervised construction of public buildings and bridges. Then in 1795, he accepted an appointment as General Agent to the Ellesmere Canal Company, because, as he later wrote, "feeling in my-

Figure 2.6 Thomas Telford (1757–1834)

self a stronger disposition for executing works of importance and magnitude [rather] than for details of house architecture I did not hesitate to accept their offer, and from that time directed my attention solely to Civil
Engineering."[14]

    There followed a series of immense projects all over Britain for
which Telford was the designer. These included the Caledonian Canal in
Scotland; the Highlands roads and bridges; the road between Shewsbury
and Holyhead including the 1826 Menai Straights Bridge, then the longest
spanning bridge in the world; as well as the Ellesmere Canal, which in-

cluded the immense Pont-y-Cysyllte aqueduct at Llangollen (see Figure 2.6). Close study of any of these projects would show Telford's great design talent, but his talent is best revealed in his iron bridges, especially his development of the wrought-iron suspension bridge and of the cast-iron arch bridge between 1795 and 1826, about the same span of time during which Watt developed his most advanced steam engines (1763 to 1792).

Heavy floods swept away the masonry bridges over the Severn River in 1795 but left the Pritchard and Darby 1779 Iron Bridge still standing. Telford, who was already skilled in masonry bridge design, recognized that an iron bridge would resist flood damage better than masonry. Because of its openness, an iron bridge did not dam the water to create the large pressure that pushed a stone bridge over. Thus, in 1795 Telford turned to the new material of industrialized iron and began three major works: the Buildwas Bridge over the Severn near Iron Bridge; the Longdon Aqueduct; and the Pont-y-Cysyllte Aqueduct.[15] Buildwas was Telford's first cast-iron arch bridge, and it can be compared to Iron Bridge in much the same way that the Newcomen engine can be compared to Watt's first successful engine design. Indeed, Telford's 130-ft span at Buildwas used about 50 percent as much iron as the 100-ft. span Iron Bridge and hence the efficiency is more than three times as great.[16] Just as with Watt, this is the difference between a craft approach and an engineering design.

Again as with Watt's earliest working engines, Telford's earliest working iron arch was still inefficient. Yet it was a great advance over Iron Bridge for three reasons. First of all, Telford carried out numerous tests on entire structural forms, not just on individual pieces. In this case, at Coal-

Figure 2.7 Iron Bridge at Coalbrookdale in England, 1779

brookdale in April 1795 he tested cast-iron arch ribs on which to base his design. Second, he followed established timber-bridge designs used in Switzerland for long spans with flat arches (rather than the semicircular arches at Iron Bridge), and third, he gave to none of the iron work any merely decorative features such as the circles between the arch and the deck at Iron Bridge. However, Buildwas has an ambiguous arch form arising from Telford's use of two arches of differing curvature.

Telford began to overcome that ambiguity in his next arch bridge design, the spectacular but unbuilt proposals for a 600-ft. span bridge over the Thames River in London. The first proposal (autumn 1800) shows remnants of arches with differing curvatures, but they no longer intersect each other in profile as at Buildwas. The design became a focus for public discussion and professional opinions. Telford made a revised design in early 1801 in which he more clearly expressed a single arch defined by X-braced panels and connected to the roadway above by a lattice of diagonals and longitudinal elements nearly parallel to the arch itself.

Some of Britain's most prestigious engineers and academics were invited to comment on the design. These included the Astronomer Royal, the Iron Masters John Wilkinson and William Reynolds, as well as James Watt and one of his most talented assistants, John Southern. It is intriguing to find that Watt disapproved of the higher ribs (the longitudinal elements nearly parallel to the arches) and advised that the diagonals be vertical—two ideas that Telford would come to in his mature arch designs later on.[17] Telford's design, although probably the best made at that time, could not have been built without substantial testing and modifications;[18] it did, however, force him to think deeply about all the major problems, and it was an essential link between the ambiguous Buildwas and the elegant forms he developed by 1812.

Telford's next iron arch proposal was equally spectacular and much more practical. It was a 500-ft. span crossing for the Menai Straits on the Holyhead road. This 1810 proposal included a construction plan to support the arch temporarily by rods that go above the bridge over temporary towers and then down to anchors at the abutments.[19] This unbuilt plan was a precursor to his 580-ft.-span Menai Straits suspension bridge of 1826 and also to his first completed arch bridge since Buildwas, the 1812 Bonar Bridge in Northern Scotland.

For Bonar, Telford again decided upon an iron arch because of flood dangers, but now the form is not ambiguous. Each of the four arches is a single, clearly visible, flat curve consisting of five identical castings 3 ft. high and in the form of four X-braced panels in each casting. This repetition led to economy. Also Telford had the entire arch erected in June

1812 at the foundry of William Hazledine at Pont-y-Cysyllte in Wales; then he had it disassembled, sent to Northern Scotland, and reassembled in the fall at Bonar, where the bridge was completed in November, a year ahead of schedule. The highway commissioners commented that the lozenge-formed connections between road and arch combined "the advantages of internal strength and external appearance."[20] Utility and beauty combined was, of course, Telford's goal.[21]

Craigellachie Bridge

Arch

q = 3.0 kips/ft. uniformly distributed load along the bridge deck*
    (one kip equals 1,000 pounds)
L = 150 ft. = arch span
d = 20 ft. = arch rise

The horizontal force at each support:

$$H = \frac{qL^2}{8d} = \frac{(3.0)(150)^2}{8(20)} = 422 \text{ kips}$$

The vertical force V (in this case, half the bridge load) at each support:

$$V = \frac{qL}{2} = \frac{(3.0)(150)}{2} = 255 \text{ kips}$$

*See *Life of Thomas Telford*, 1854, p. 685. The ironwork weighs 2 kips/ft., and we assume an additional load of 1 kip/ft.

Telford's next cast-iron arch was the Craigellachie Bridge built in 1814, also in Northern Scotland, over the Spey River near Elgin. Almost a replica of the Bonar Bridge, the Craigellachie still stands today although partly rebuilt in 1965; the Bonar Bridge was destroyed in a flood about 1892 but not because of any problems with the iron work. For the Craigellachie form, as pure as that at Bonar, Telford did design stone-turreted towers on the approaches, an architectural embellishment that does not fit with the span but is far enough away from it not to disturb the pure engineering form.

In 1824 Telford designed a 170-ft. span, cast-iron, arch bridge over the Severn at Twekesbury for which he oriented the lozenges (the X-shaped connections between deck and arch) vertically instead of radially as he had done in his earlier Scottish bridges. He repeated this more rational detail for the 1829 design at Galton, his last cast-iron arch bridge. In those last works, we can see the modern arch bridge form almost fully prefigured. The longest spanning arch at present (in 1996) is the 1,700-ft. span over the New River in West Virginia. It has flat arches formed by braced panels and is connected to the deck above by thin vertically oriented elements.[22]

Telford and Watt

Having spoken of Telford's designs themselves, we must say something about how he worked in order to compare him more carefully to Watt. Whereas Watt had the instincts of a scientist, which helped him to ask basic questions about engines, Telford had the instincts of an artist, which led him to think of his designs both in technical and in aesthetic terms. In both cases their designs were more impressive because each man looked beyond mere precedent and asked more fundamental questions about his works. But the questions each asked were quite different.

For Watt the questions had to do with such ideas as latent heat, the expansion of steam, and even the pollution from fuel burning, whereas for Telford the questions had to do with local foundation conditions, techniques of construction, and the visual image of a completed work. Watt's questions were ideal for a sedentary and reflective personality; Telford's questions required constant traveling and detailed localized study. Watt's mind dealt with things in motion, but he worked mainly at the laboratory in his home. Telford's mind centered on things designed to stand still, but he, therefore, had to be continually in motion.

Watt needed his entrepreneurial partner, Boulton, to create the environment in which he could invent; Telford had no partner and what he

needed was rather a continuing friendship with financial and political figures who decided on commissions for canals, harbors, roadways, and
bridges. Watt was introspective and had little talent for managing people
or accounts. Telford was outgoing, humorous, and gifted in organizing a
building site or laying out an immense stretch of canal or road.

Telford was not a private entrepreneur with a fixed factory complex to run like Boulton; rather he was more a manager of ad hoc construction sites and shifting local labor forces. Indeed, he rarely was what
we would call the builder or contractor on his works. Rather he represented the owner and supervised the builder, who was often one of his
hand-picked colleagues. Watt with Boulton was part of a relatively large
industrial firm; Telford acted mostly alone and in combination with others
who changed from project to project. Telford often invested in his projects
and he did have a small group of assistants, mostly Scots, who worked
continually with him, but his focus was always on specific local activities
rather than on the running of a centralized factory complex.

Often modern engineering is thought to be defined by industry
and machines. Even the name *industrial revolution* connotes the identifi-
cation of modern technology with machine industry. However, in Telford
and Watt, we find two sides to technology—structures and machines—
each essential to the industrial revolution and each still essential today.
The canals, roads, bridges, harbors, and later the railways made modern
industry possible, but those modern structures themselves would have
been impossible without the steam engine and all of its later factory developments.

Design and New Ideas

In the cases of Watt and Telford we can identify a series of new
ideas coming from the stimulus of design. The design problems of the
steam engine stimulated Watt to new ideas of measurement: the idea of
horsepower and of the pressure-volume indicator card attached to his engines. More fundamentally, by the mid-nineteenth century it stimulated a
new branch of natural science, thermodynamics, that was a direct result of
the steam engine.[23]

In a similar manner, Telford's studies and tests for the Menai
Straits Bridge stimulated the more analytically minded to develop a new
theory of structure. For example, Telford's first design of 1820 stimulated
Davies Gilbert, a Holyhead road commissioner (and later to become president of the Royal Society), to develop a mathematical theory for suspension bridges that he later published.[24] The design again stimulated the

search for a theory and not the reverse. Indeed, the fame of Menai travelled quickly to France and led C. L. M. H. Navier to visit Britain, study
the bridge, and eventually write the first systematic treatise on such
works.[25] Just as other machine designers were patenting new designs con-
temporaneously with Watt, so other structural engineers were building
suspension bridges at the same time as Telford. Watt and Telford are the

Menai Straits Bridge

Suspension Span

$q = 3.24$ kips/ft.-uniformly distributed dead
and live load along the bridge deck*

$L = 580$ ft. = bridge deck span between towers

$d = 43$ ft. = cable sag

The horizontal force at the top of each tower (due to the main span loading):

$$H = \frac{qL^2}{8d} = \frac{(3.24)(580)^2}{8(43)} = 3,168 \text{ kips}$$

The vertical force V at the top of each tower (due to the main span loading):

$$V = \frac{qL}{2} = \frac{(3.24)(580)}{2} = 940 \text{ kips}$$

*See R. Paxton, *Influence of Thomas Telford*, 1975, pp. 390–391.

two leading figures in fields that were expanding with unprecedented speed and success.

Along with new ideas on steam and suspension came the new physical facts of rotative engines in textile mills and fixed all weather transportation links from Britain's major cities out to the far reaches of the island—Holyhead to the west and Dornoch to the north. These physical facts created two central social ideas of the nineteenth century: the organization of a productive process and the linking of city and country by a transportation network.

New engineering designs, therefore, stimulated ideas about society that transcended responses to *individual* objects (machines or structures) and began to include *systems* of objects (processes and networks). This grander scale of thinking was foreshadowed in Boulton's famous 1769 letter to Watt, in which the entrepreneur described a system of manufacturing with excellent tools, precision work, economical prices, and a market consisting of "all the world."[26] The idea was, of course, to produce the greatest amount of work for the least energy and to secure thereby the greatest profit through repetitive action—a good description of both the steam engine itself and of the manufacturing process of steam engines or of textiles.

Finally, the new physical realities of structures and machines, combined with the new social ideas of networks and processes, stimulated a set of ideas about modern society itself. One group of people, impressed by the new power and great extent of modern engineering, saw society improving continuously and with the prospect for increasing improvement in the future. As one commentator put it in speaking about nineteenth-century technology, "The law of progress is immortal, much as progress itself is infinite."[27]

Set against that optimistic view was an opposite one that concluded that the industrial revolution brought with it only disruption, horror, and oppression. Both optimists and pessimists agreed that there was change and that the change was due to modern engineering. They disagreed only in their view of its direction: upward to Utopia or downward into hell. There was also a third view that combined the first two and believed history to be cyclical: Things got better, then they got worse, and so on. None of these ideas was new; each had been articulated well before the industrial revolution.[28] What was new was the strong physical imagery of modern engineering: factories, railroads, and crowded cities, along with huge structures and powerful machines, having new and unprecedented forms such as the Eiffel Tower and the Brooklyn Bridge, the Corliss Engine of 1876, and the Halls of Machines in Chicago (1893) and in Paris

(1889 and 1900). The idea of change as progress suddenly acquired an immense physical reality. What had been the speculation of philosophers before 1800 became the talk of the masses by 1900. Progress gained a popular constituency that debates these ideas even more vehemently in the late twentieth century.

Is engineering the embodiment of the idea of progress? Is there an unalterable drive for continual change and advance toward ever more efficient structures and machines? Will the problems seemingly brought on by each new technological advance be solved by the next advance or merely made worse? We find the key to answering these questions in the idea of design itself.

Viewed from the perspective of invention, engineering surely provides an image of progress. By definition, an invention must be new, and those that have influence are normally aimed at improving efficiency. Design as invention indeed moves inexorably on and drives industry from piecework to mass production; from manned assembly lines to robotics; from rooms full of engineers on drawing boards to lonely computers producing numbers, drawings, and machine parts. This mechanized life is the direct result of design as invention as it was brought into the commercial world by Boulton and Watt 200 years ago.

But there is another side to modern technology, another set of physical images in which design is art as first brought into the public domain by Thomas Telford. In what sense, for example, is a twentieth-century steel bridge more efficient than one of Telford's iron bridges? The distinguished engineer Sir Alexander Gibb, in his 1935 biography of Telford, had this to say: "In his iron bridges, Telford so closely studied the elimination of weight of material that they have been refined to a delicacy that has perhaps never been equaled. Such a bridge as Craigellachie Bridge surely comes near to the ideal."[29]

Of course, we would not build today a cast-iron arch of 150 ft. in span, but are our steel bridges designed as efficiently and will they last as well as Telford's? We cannot say, but we do know that a staggeringly large number of twentieth-century bridges have not lasted as well as Craigellachie and that very few are as elegantly designed. Whereas by the end of the nineteenth century not a single Watt engine was left in service, nearly every one of Telford's bridges was still serving.[30] Progress in bridges, therefore, has a different meaning from progress in steam engines. Clearly, twentieth-century bridge designers do try to improve on their predecessors' designs, but not in the same way as do machine designers.

Telford and some of the most talented succeeding structural designers viewed themselves as artists working in a new medium: iron, steel, and reinforced concrete, materials made possible by the industrial revolution. In his 1812 treatise on bridges, Telford defined the new art form as arising out of the disciplines of efficiency (safe performance with minimum materials) and economy (utility with minimum cost) but marked off by the search, within these disciplines, for elegance of form and detail. This new art form, he recognized, was parallel to but independent of architecture and required detailed engineering experience and training.[31]

Because the best structures are works of art, we need to think in terms of art history to define progress in bridges, towers, long-span vaults, and other designs by structural engineers. In painting and sculpture there is a kind of progress, as the best young artists do not want to copy their elders. But we surely cannot say that Cézanne is better than Turner; we can only say that each was among the best of his era. In the same way the spectacular wrought-iron railway arch bridges of Gustave Eiffel are the best of their era (1860–1890), just as Telford's cast-iron roadway arch bridges were the best of his. In art there is change, but there is not necessarily progress, whereas in science we can clearly say there is progress in the sense that the achievements are cumulative. Einstein knew more than Newton and Newton more than Kepler. Thus, again, machine advances are more like those in science whereas structural advances are more like those in art.

Although one can separate art from invention and imply that the former explains structures and the latter machines, in fact the two ideas are connected in complex ways. More than anyone else in the early years of U.S. independence, Robert Fulton (1765–1815) illustrates the intimate connection between art and invention both literally in his two-part career and figuratively in his principal achievement, the steamboat. Favorable conditions played a crucial role: U.S. independence and a vast continent as yet unsettled.

3 Fulton's Steamboat and the Mississippi

Fitch and Rumsey

It has been argued that "the steamboat was the first great American contribution to modern technology."[1] The first successful steamboat designer was John Fitch (1743–1798), who got a boat about 30 ft. long to go against the wind and tide on the Delaware River at Philadelphia on August 22, 1787. By 1790 he was able to establish a service between Philadelphia and Burlington but it was never a commercial success.[2]

Fitch's principal rival in the 1780s was James Rumsey (1743–1792) of Maryland, who had strong connections with high political figures including George Washington, Thomas Jefferson, and Benjamin Franklin. Fitch's apparent successes pushed Rumsey in 1787 to make two trial runs of his own steamboat design on the Potomac River. He succeeded in going three miles per hour in August of 1787 but the boiler failed partly. After the boiler was repaired he tried again on December 3 and once more on December 11, 1787, but again Rumsey failed. It was always in the boiler or pipes that cracks appeared and destroyed the engine power.

Rumsey had tried a simplified jet-propulsion idea whereas Fitch had used a set of vertical paddles imitating, in principle, galley slaves. Both inventors used reciprocating engines that they designed themselves.[3]

Fitch and Rumsey both tried to get local monopolies to fend off each other and establish priority of invention. Finally, on April 10, 1790, the federal government entered the scene when the Congress passed the first U.S. patent law. It was intended primarily to deal with steamboat claims. The first commissioners were Jefferson, then secretary of state, Henry Knox, secretary of war, and Edmond Randolph, attorney general. Thus the steamboat entered politics and began a transformation of government that would culminate more than 60 years later in the first direct intervention of the U.S. government into the affairs of private industry (the steamboat companies).

No clear patent priority resulted in the Fitch-Rumsey competition. Fitch kept trying but his difficult personality and alcoholism lost him financial backers, and by 1793 he gave up, dying five years later. Rumsey, always an effective salesman, kept his hopes alive with numerous schemes, including trying to circumvent the patent of James Watt, but he, too, did not last far into the 1790s and died on December 18, 1792, while writing a report on new ideas for steamboats. Thus, by 1793 both Fitch and Rumsey had left the stage and a third figure appeared with a quite different approach. It was in June of 1793 that Robert Fulton "for the first documented time tried his hand with the steamboat."[4]

Robert Fulton: Artist and Designer

Fulton, who had gone to London in 1787 to study painting with Benjamin West, began about 1793 to study canals, machines, and civil engineering. In 1796 he published the *Treatise on Canal Navigation*,[5] which included a series of plates made by Fulton himself illustrating his artistic talent and his growing interest in boat design, the design of machinery, and the design of aqueducts. The text of the book shows Fulton's entrepreneurial motivations as well as his strong engineering bent.

He was not successful in promoting canals so he turned more and more to boats. In 1803 he built a steamboat in France, which provided him with enough confidence to proceed with designs for one in the United States. Fulton's steamboat trial of August 9, 1803, on the Seine did not achieve the 16 miles per hour (mph) he had calculated. Rather it made barely 4 mph and his principal backer was unhappy as was France's leading personality, who had sent a representative to the trial. Upon hearing of the greatly reduced speed, Napoleon is supposed to have stated that "many charlatans or impostors . . . have no other end but to make money. That American is one of them. Don't speak to me about him any further."[6]

Unlike Fitch and other earlier steamboat designers, Fulton did not try to design his own engine but rather used a Watt engine. In fact, Fulton tried as much as possible to use already available ideas and components. His goal was to design a successful steamboat, not to make new engine designs.[7]

Fulton had a financial backer, the wealthy Robert R. Livingston, who owned an imposing estate at Clermont, New York, on the Hudson River. Livingston secured from a political body (the New York State Legislature) the promise of a monopoly. The legislature granted to him and

Figure 3.1 Robert Fulton (1765–1815)

Fulton exclusive rights for 20 years to the New York waters so long as they could operate a 20-ton boat at the rate of 4 mph.

On August 17, 1807, at 1:00 P.M. Fulton's "North River Steamboat" (soon to be called the *Clermont*) went up the Hudson River (known in New York City as the North River) to Clermont. The trip took 24 hours.

Figure 3.2 Iron aqueduct of 1796 designed by Robert Fulton

The boat then proceeded to Albany. The total running time was 32 hours over a distance of 150 miles, thus exceeding the speed of 4 mph.[8] Fulton's success came from both his own design skill and his partner's political power.

The design skill of Fulton included a mode of thinking akin to the visual arts. "The mind of the individual inventor or projector was the ultimate key. . . . [T]he men who emerged as the most effective in developing designs of complete steamboats based upon individual and unique combinations of a complex of elements all enjoyed a capacity for spatial thinking."[9] Fulton himself understood clearly the significance of this kind of thinking. "The mechanic," meaning in today's language the mechanical engineer, he wrote, "should sit down among levers, screws, wedges, wheels, etc. like a poet among letters of the alphabet, considering them as the exhibition of his thoughts, in which a new arrangement transmits a new Idea to the world."[10]

As this quotation implies, Fulton was both a designer and a promoter. He tried to convince people with money and political power to sponsor his research, and he succeeded remarkably well. His steamboat success came in part because of his abilities and in part because of the economic opportunity that made powered transportation over water highly desirable in the new nation. This "economic pull" was not as strong in either Britain or France, where there were not the long navigable rivers reaching into the wide expanses of territory.

Fulton's Patent Design

On January 1, 1809, in Washington, D.C., Robert Fulton wrote out his patent application for the steamboat. By describing Fulton's patent, we can gain insight into the origins of innovations. After first describing his method of connecting a Boulton and Watt engine to the paddlewheels, Fulton states that his design depends:

> First, on an accurate knowledge of her total resistance while running 1, 2, 3, 4, 5, or 6 miles an hour in still water; second, on a knowledge of the diameter of the cylinder, strength of the steam and velocity of the piston to overcome the resistance of a given boat while running 1, 2, 3, 4, 5, or 6 miles an hour in still water; third, on a knowledge of the square feet or inches which each propeller should have and the velocity it should run to drive a given boat 1, 2, 3, 4, 5 or 6 miles an hour through still water.
>
> It is a knowledge of these proportions and velocities which is the most important part of my discovery on the improvement of steam boats.[11]

Next he proceeded to put those three basic questions into numerical calculations.

First, he presented a table of numbers taken from the extensive English experiments in the 1790s by Mark Beaufoy. This table recorded experimental results of drag on differently shaped objects going through water at speeds from 1 to 6 mph. Second, Fulton showed how to relate these drag values to the characteristics of a steam engine to give enough power to move the boat, and third, he gave the calculations needed to determine the size of the paddles, which he called the propellers.

He took a boat 154 ft. long, 18 ft. wide, drawing 2 ft. of water, and having an angle of the bow and stern (observed from above) of 60 degrees. From these dimensions he calculated a total *drag* on the boat of 918

Figure 3.3 1809 patent drawing for Fulton's steamboat

Figure 3.4a Sketch of *North River* steamboat from Fulton's notebook

Figure 3.4b Drawing of *North River* steamboat's machinery

lbs. when going 4 mph, the minimum speed that had been necessary in 1807 to secure the monopoly franchise legislated by New York State.

In theory for a given vehicle (boat, locomotive, car, plane, etc.), the drag should be directly proportional to V^2. Fulton's patent calculations show this to be close to the case.[12]

Fulton now had to show that the power of his engine was enough to drive the boat 4 mph. The paddlewheels must, of course, go faster than 4 mph and must give a thrust, T, against the water of 918 lbs. to overcome the drag. Fulton assumed that the paddles would travel 8 mph, or 704 ft. per min[13]. The power of the wheel is then the product of its velocity, V, and its thrust against the water. Because James Watt defined horsepower as 33,000 ft.-lbs. per min., it may be defined as Hp = TV/33,000.

Therefore, the power needed to push the boat through the water was 704 × 918 = 646,000 ft.-lbs. per min., and the steam engine had to provide that power. Power, as we see, is equivalent to lifting a weight (or pushing an object against a resisting force) for a given distance in a specific length of time.

Fulton had purchased a Boulton & Watt 20-horsepower steam engine, modified by Fulton himself, which was already in New York when he arrived back from England on December 13, 1806. Thus the power required by Fulton's design was 646,000/33,000 = 19.6 horsepower, or just about what he had provided by his Watt engine.

To make the paddles go 704 ft. per min., Fulton designed two paddlewheels 14 ft. in diameter, one on either side of the boat, and calculated that they must travel at a speed of 16 revolutions a minute so that the paddles, moving along the wheel's circumference, would have a velocity $V = 16 \times 14\pi = 704$ ft. per min. (fpm). He was concerned to get the ideas right and to describe them numerically, but absolute precision made no sense because of the considerable uncertainties in the experimental data on which the 918 lbs. were based and in the wind and water conditions to be faced in an actual full-scale run. Indeed, Fulton provided his boat with sails in case of heavy winds.

Fulton's Patent: The Formulas

In his patent Fulton claims to have been the first to publish "exact or mathematical principles . . . to guide artisans to success in works of this kind." His leading example was a boat 154 ft. long, 18 ft. wide, drawing 2 ft. of water, and powered by a steam engine purchased from Boulton and Watt. Fulton's mathematical principles follow:

1. Water resistance to boat movement = drag (D),

    $D = 918$ lbs. at 4 mph

2. Paddle power (P_P) to overcome this water resistance is supplied by engine power (P_E).

    $P_P = 19.5$ Hp

    $P_E = 20.8$ Hp

3. Paddle wheel* area (A_P) and velocity (V_P) to move his boat:

    $V_P = 8$ mph
    $A_P = 17.7$ sq. ft.

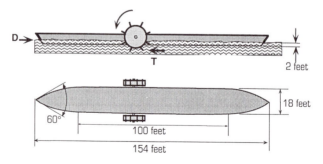

1809—Boat in Fulton patent

*Fulton called his paddle wheel a "propeller."

Fulton's last calculation aimed at dimensioning the paddles, and for that he again relied on Beaufoy's test data, which showed that a 1 sq. ft. flat plate pushed against water running at 4 mph with a force of 51.95 lbs.[14] Therefore, the total force of 918 lbs. required an area of paddle equal to 918/51.95 = 17.7 sq. ft., or 8.85 sq. ft. on each wheel (again Fulton used an approximate value of 8.75 sq. ft.). He concluded that "by this example all necessary calculations may be made."[15]

We observe here an early example of modern engineering design. First, the dimensions came from calculations based upon scientific study of objects in moving water and of steam engines in service. Second, the calculation methods themselves are correct even by the standards of today although engineers now use new experimental data and more so-

Fulton's Drag Calculations

Drag (D) is defined as the *water pressure* (p) on the *wetted surface area* (A) of the hull, multiplied by the *drag coefficient* (C_D), and thus the formula is:

$$D = p A C_D$$

Water pressure (p) is the force in pounds on an area of 1 sq. ft. caused by water with a mass density (ρ) calculated by dividing the density of water ($w = 62.4$ lbs./cu. ft.) by the acceleration of gravity ($g = 32.2$ ft./sec.^{2})

$$p = \frac{1}{2}\rho V^2 = \frac{1}{2}\left(\frac{62.4}{32.2}\right)V^2 = 0.97V^2$$

which for $V = 4$ mph = 5.87 ft./sec., gives

$$p = 0.97(5.87)^2 = \textbf{33.4 lbs./sq. ft.}$$

The *wetted area* (A) Fulton calculated for his boat was 3,048 sq. ft. For $V = 4$ mph Fulton calculated the drag to be 918 lbs. based upon data he found in Beaufoy's report. Fulton's result would lead to a *drag coefficient*: $C_D = 0.009$.

$$D = 33.4\ (3,048)\ 0.009 = \textbf{918 lbs.}$$

Using present-day data, we would find a much lower $C_D \approx 0.0022$ (see note 15) and hence a drag of about 225 lbs.

phisticated equations. Third, in the early stages of development there is no need for calculation precision so long as the basic formulas are correct and so long as full-scale tests confirm the results. Fulton had already shown in August 1807 that his boat would go 4 mph and even somewhat faster. Indeed in his patent calculations he far overestimated the drag on his boat,

Fulton's Power Calculations

Paddle power (P_P) defined by the force on the paddles (thrust = T) and the velocity of the paddles (V_P):

$$P_P = \frac{TV_p}{33,000}$$

Thrust (T) must equal the drag to move the boat

$T = D =$ **918 lbs.**

Paddle velocity (V_P) must be greater than the boat velocity to move the boat, and Fulton chose:

$V_P = V + V_T = 4 + 4 = 8$ mph = **704 ft./min.**

where V_T is the velocity that provides the thrust.

The *paddle power* is

$$P_P = \frac{918(704)}{33,000} = \textbf{19.6 Hp}$$

Engine power (P_E) is defined by the cylinder pressure ($P = 10$ lbs. per sq. in., or psi), piston area ($A = 27^2 (\pi/4) = 573$ sq. in.), stroke length ($L = 4$ ft.), and piston speed ($N = 30$ strokes per min.):

$$P_E = \frac{10(4)(573)(30)}{33,000} = \textbf{20.8 Hp}$$

This is enough to supply the needed paddle power of 19.6 Hp.*
Had Fulton chosen a small paddle velocity he would have had to increase the paddle size (see next figure).

*There must have been significant losses in transferring cylinder power to paddle power. Since Fulton far overestimated the drag, he inadvertently compensated for the lower paddle power (see note 15).

and he also way underestimated the power available at the paddles. These were compensatory errors and escaped his notice because the boat performed as he predicted. Fulton's method was modern, even if his computations were not.

Fulton's Paddlewheel Calculations

Paddle size (A_P) depends upon the *thrust* (T), the *paddle velocity* (V_P), the *boat velocity* (V), the *water pressure* (p_T) corresponding to the *thrust velocity* ($V_T = V_P - V$), and the *drag coefficient* (C_D) for the paddles:

$T = p_T(A_P)C_D$

Thrust (T) must equal the drag (D):

$T = D = \textbf{918 lbs.}$

Thrust velocity (V_T) must be $V_T = V_P - V$

Fulton chose the $V_P = 8$ mph so that

$V_T = 8 - 4 = 4$ mph$= \textbf{5.87 ft./sec.}$

Thrust pressure (p_T) is the water pressure due to thrust velocity:

$$p_T = \frac{1}{2}\rho V_T^2 = 0.97(5.87)^2 = \textbf{33.4 lbs./sq. ft.}$$

Drag coefficient for the paddle was based on Beaufoy's report, which gave data on plates moved against water pressure. For a thrust velocity of 4 mph, Beaufoy gave a value of 51.95 lbs. for a 1-ft. square plate. Thus, Fulton could calculate the total paddle area A_P needed to provide 918 lbs. of thrust as follows:

$$A_P = \frac{918}{51.95} = \textbf{17.7 sq. ft.}$$

This amounted to a drag coefficient of

$$C_D = \frac{T}{p_T A_P} = \frac{918}{33.4(17.7)} = \textbf{1.55}$$

which is higher than engineers would use now (see note 15).

Origins of Innovations

The search for origins of events and objects is vital both to historians and to designers. It is intrinsically of great human interest to search out the past and then find a source; it is equally compelling to discover a general truth that might then be applied to change the future course of events or the design of objects. If people can agree on how innovation arose, then by the proper investment of public or private funds they can bring even more useful innovations to maturity and even more rapidly than in the past. The problem is, of course, to agree on the historical record, and that record is always more complex than any simple theory of innovation can encompass. It is possible, nonetheless, to identify three competing ideas about the origins of technological innovation: one, of innovation as a consequence of applied science; two, of innovation as a response to political and economic forces; and three, of innovation as the result of individual genius. We can see to what extent each of these ideas is true of Robert Fulton.

Innovation as Applied Science

The first idea stresses the prior vision of the natural scientist, whose discoveries then stimulate technical people to new applications. This applied-science theory of innovation was directly responsible for the creation after World War II of the National Science Foundation (NSF). Vannevar Bush, a distinguished electrical engineer, articulated this view in a 1945 report to President Harry Truman when he wrote:

> Basic research leads to new knowledge. It provides scientific capital. It creates the fund from which the practical applications of knowledge must be drawn. New products and new processes do not appear full-grown. They are founded on new principles and new conceptions, which in turn are painstakingly developed by research in the purest realms of science.
>
> Today it is truer than ever that basic research is the pacemaker of technological progress. In the nineteenth century, Yankee mechanical ingenuity, building largely upon the basic discoveries of European scientists, could greatly advance the technical arts.[16]

Yankee mechanical ingenuity as exemplified by Robert Fulton, according to this view, must have built largely on the basic discoveries of European scientists, who had worked "in the purest realms of science." What scientific discoveries and which scientists influenced Robert Fulton? Very few. He certainly spent enough time in England and France to absorb new ideas from science. The record of the 1803 design for Fulton's first successful steamboat trial in Paris shows his own imagination and his own ideas in charge even though he certainly relied on earlier research, earlier failed designs, and collaboration with numerous experts.[17] He got the side paddlewheel idea from earlier French designs, he used a Watt engine, and he experimented with new shapes for the hull, measuring "reduced resistance and increased velocities." He based his calculations for the hull on prior French and British research on ship resistances, such as Beaufoy, but he relied directly on his own experiments for the design. He was, therefore, fully aware of prior designs (mostly failures) and of prior research (mostly engineering research), and he even used results from French mathematician Charles Bossut, who had himself collaborated with engineers.

It is clear that Fulton conceived of his own design but also that he looked for help wherever he could find it. He did not, however, build largely on discoveries of European scientists. He was a prototypical engineering designer; that is, he derived primary stimulus from studying and observing earlier engineering designs and recognizing their defects. Those difficulties then guided him in his search for a better overall design. He attacked each major part of the new design by using as much as possible proven ideas or ones he could test for himself. The scientific theories of hydraulics and thermodynamics lagged very far behind Fulton's design. His works, together with those of Watt and others, would in fact be the direct stimulus to a new generation of European scientists starting with Nicolas Léonard Sadi Carnot (1796–1832), who would develop theories out of the practical innovations of engineers. Thus the applied science of innovation is sometimes turned upside down when it is applied to early-nineteenth-century objects of Yankee mechanical ingenuity.

Innovation as a Social Process

If the applied-science view cannot fully explain Fulton's success, then perhaps it can be understood better as the product of two types of social forces. It was Louis Hunter, America's leading historian of steamboats, who said: "The Steamboat was the first great American contribution to modern technology . . . and in the popular history of American technology Robert Fulton occupies a place . . . [on] a high pedestal . . . comparable to that of James Watt in England." Hunter goes on to argue that Fulton is not to be credited with that innovation but rather that he and a few others "would assume a quite modest position beside the collective contributions of scores of master mechanics, shop carpenters, and shop foremen in whose hands the detailed work of construction, adaptation, and innovation largely rested."[18]

Hunter sees innovation here as a social process in the sense of a society of craftsmen or engineers who as a profession push forward through the waters of the unknown to arrive at the haven of new designs. This is the antiheroic theory of innovation as a reaction to the "great man" theory of history.

Another side to that reaction comes from the recognition that neither Watt nor Fulton could have succeeded without the forces of politics and economics pushing them along. Without financial backing, without political monopolies, and with no natural market, their ideas would have gone nowhere. In short, Watt and Fulton are names given as convenient shorthand for two types of social processes: one from within a technical profession and the other from within a technologically developing society. Instead of seeking innovation from the scientific community, we are directed now both to seek it within the technical profession collectively and to call for it from our political leaders in government—not the National Science Foundation but the Engineering Founders Societies and the Congress. But to show that innovation arises primarily from such places is even more difficult to demonstrate than that it comes from the purest realms of science. The Congress and the British Parliament did react to steam engines and to steamboats, and they did at times frustrate and at times facilitate advances, but they did not stimulate design ideas. The professions of mechanics, carpenters, and foremen certainly were essential to the carrying out of the designs of Watt and Fulton, but neither man was centrally stimulated by them.

Innovation by Individual Genius

It is not necessary to abandon the belief that applied science and social processes stimulate innovation to recognize the central role of individuals in engineering advances. Probably no one made any greater individual contribution to the industrial revolution than James Watt. In the same way Fulton deserves recognition as the pioneer in commercially successful steamboat designs. His command of the engineering calculations and his motive to see his ideas realized commercially were basic, but his personality was also a major factor. He was persistent and often persuasive. He was able to attract the support of a powerful political figure such as Livingston.[19]

From this brief outline of Fulton's design and the origins of the steamboat, we see the importance of existing scientific data and ideas (especially on drag of boats in water) as well as of existing engineering designs (especially the steam engine). We see the role of individual engineers and politicians in achieving an innovation, all of which illustrates the interrelated roles of applied science, social processes, and individual people in the building of a technological society.

The Scientific Perspective on Steamboats

When the steam engine moved from the mine or factory to the boat, the design problem changed from being a stationary power plant to powering a moving vehicle. Fulton succeeded where Fitch and Rumsey failed partly because he kept the Watt stationary engine. But he paid a big price because much of the engine's power went to carry the heavy engine itself, and its great size subtracted usable space onboard. The new problem, evident once the *Clermont* succeeded, was to reduce drastically the engine weight and size. The horsepower can only be kept the same when the engine size and weight are radically dropped if the steam pressure or the piston speed are increased proportionally.

To put the issue numerically, we recall the power formula, *PLAN*/33,000, and we see clearly that the heavy cylinder, expressed by *LA*, can only be reduced if *P* and *N* are correspondingly increased. In theory it was easy to increase P. Watt recognized this clearly but he felt that the metal technology was not reliable enough to permit high internal boiler pressures.

The first person in the United States to design commercially successful high-pressure steam engines was Oliver Evans (1755–1819), who built his first such engine in Philadelphia in 1801. He got a patent in 1804 and for a time was the only manufacturer of high-pressure steam engines in the United States. In 1815 his patent, like Watt's patent of 40 years earlier, was extended by an act of Congress.[20] In 1816 Evans designed an engine for the steamboat *Aetna* to run between Philadelphia and Wilmington, Delaware. Later on the boat made the run between New Brunswick and New York. It then had three boilers and was reputed to work at pressures as high as 149 psi (as opposed to the 7.5 psi found often in atmospheric engines whose theoretical maximum would be about 14.7 psi). It was this increase in pressure of about 20 times that allowed Evans to design much smaller engines, which meant that for equal horsepower the cylinder volume need be only one-twentieth of that required for an engine operating with 7.5 psi.

Figure 3.5 Explosion of the *Cincinnati*

However, a set of new engineering problems arose. The primary one appeared in the boiler because of the need to contain in an iron vessel the very much higher pressure. It is certainly clear that the strength of the vessel must increase by a factor of 20 if the load (internal pressure) increases by that factor.

This high pressure immediately led to the central scientific difficulty with steamboats: the explosion of the boilers. In May of 1824, while in New York Harbor, one of the *Aetna*'s three wrought-iron boilers burst, killing five or six people and injuring many others. One explanation of the failure was that a longitudinal crack had started at the rivet connection directly above the frame. Ring stresses caused these cracks to enlarge and led to failure. The rivet holes weakened the boiler plate, and also it was frequent practice in order to get more power to shut down the safety valves on the boiler to secure greater pressure; this of course led to greater stresses.[21]

To understand the boiler problem better, we look at the ring stresses that develop in a cylindrical shell under high internal (steam) pressure. We can take a simplified design for the *Aetna* boiler by assuming it to be 12.5 ft. long (150 in.), 2.75 ft. in diameter (33 in.), 0.4 in. thick, and with circular plates closing each end.[22]

Because the pressure used by Evans was about 149 psi, the axial stress in the cylinder wall (see figure) is

$$f_2 = \frac{Pr}{2h} = 3{,}073 \text{ psi}$$

This represents a stress in the cylinder in the direction of its axis, hence the term *axial stress*. The circumferential stress is

$$f_1 = \frac{Pr}{h} = 6{,}146 \text{ psi}$$

Boiler Axial Stresses

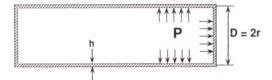

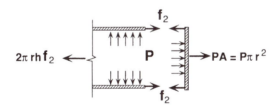

The force against the circular plates with area of $A = \pi r^2$ is PA. The connection between the plate and the cylinder end that resists that force has an area equal to the circumference of the cylinder end and the cylinder thickness (h), or $2\pi rh$. The stress f_2 pulling against the cylinder end times that area will give the force required to resist PA:

$$P \pi r^2 = 2\pi r h f_2$$

For this boiler the thickness $h = 0.4$ in., the pressure $P = 149$ psi, and the radius of the cylinder $r = 16.5$ in., therefore

$$f_2 = \frac{Pr}{2h} = \frac{149(16.5)}{2(0.4)} = 3{,}073 \text{ psi}$$

Boiler Circumferential Stresses

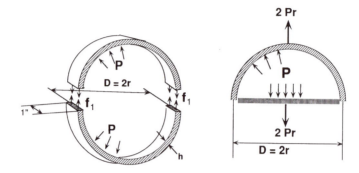

Imagine a 1-in. slice of half the cylinder to be closed by a plate such that the vertical, upward force due to pressure P will be equal to a downward force $2Pr$.

But this force is taken by a stress in the two walls of the cylindrical slice, whose combined areas are $A = 2 \times 1 \times h = 2h$ so that $2Pr = 2hf_1$, from which $f_1 = P\ r/h$ in the direction of the circumference. Therefore, this stress in the *Aetna* boiler was, with $P = 149$ psi, $r = 16.5$ inches, and $h = 0.4$ inches,

$$f_1 = \frac{Pr}{h} = \frac{149(16.5)}{0.4} = 6{,}146 \text{ psi}$$

or just twice the axial stress.[23] Therefore we should expect longitudinal cracks to be critical in boilers.[24]

In spite of these accidents, steamboats rapidly opened up the western waters (the Mississippi and the Ohio) and became an immensely popular means of travel.

The Social Perspective: Fulton and Shreve

Successful steamboat designers like Fulton benefitted from two social factors: first, the grant either of a monopoly or of a patent, and second, the American attitude against government regulation and control. This is a paradox. On the one hand Fulton needed the governmental protection of a monopoly on the steamboat service around New York Harbor to justify the initial investments, but on the other hand he benefitted by the freedom to design his boats as he wished to gain a maximum profit.

Just as with Watt, who had operated under a monopoly, Fulton took a conservative view of the engine design. Indeed he took Watt's design and used low-pressure steam, the separate condenser, and the kinematic system of converting reciprocating motion into circular motion by means of a walking beam. These three ideas had resulted in reliable and fuel-saving designs but they also were heavy and awkward. Watt's designs had proven effective for fixed power plants: pumps and factory power. They were, however, not so well suited for mobile power plants where lightweight and compact construction were initially of greater importance than reliability and fuel economy.

The success of Robert Fulton around New York encouraged him to try to secure a monopoly for the steamboat routes on the Mississippi River, already the major trading route in the western part of the United States. The Louisiana Purchase of 1803 stabilized the Mississippi Valley politically and the economic pressure from cities on the Ohio River created a vigorous competition among steam-engine and boat builders to capture the trade routes from Pittsburgh to New Orleans. Fulton's partner, Robert Livingston, along with James Monroe, negotiated and signed the $15 million Louisiana Purchase. From that time on Fulton saw the Mississippi as the place to operate; the Hudson River was only a place to begin.

Following Fulton's 1807 success on the Hudson River, he and his backers built a steamboat called the *New Orleans* (371 tons) that began to operate on the lower Mississippi River in 1811. A second boat, the *Vesuvius*, began operation in 1814 just as the *New Orleans* was sunk in an accident. The *Vesuvius* burned and sank in 1816, the year the *New Orleans* was rebuilt and thence successfully plied the 250-mile New Orleans–Natchez trade on the lower Mississippi. Fulton had died in 1815 but not before proving the value of his operations in the West despite the disasters that befell his boats.

The first *New Orleans* during its first year cleared $20,000 for its owners "above expenses, repairs, and interest on investment on a property valued at $40,000."[25] This remarkable return naturally encouraged others to compete with Fulton's group and, in addition, made the risks of sinking seem worth taking.

The principal figure in this competition was Henry Miller Shreve (for whom Shreveport was named), who had considerable experience, as a river captain and boat owner, going between Pittsburgh and New Orleans. In 1814 Shreve joined with a group of men at Brownsville on the Monongahela River not far from Pittsburgh to challenge the Fulton

Figure 3.6 Henry Miller Shreve (1785–1851), innovator of the
      steamboat

group. Already in 1813 the group had built a very small steamboat, the
Comet (25 tons), which was a failure. But the next year, their *Despatch*
(25 tons) succeeded somewhat, and their larger *Enterprise* (75 tons) became quite successful.

Shreve was part owner and captain of the *Enterprise*, which
gained fame for making the first trip from New Orleans to Louisville. In

Figure 3.7 The *Washington* steamboat of Henry Shreve

1815 Shreve made this trip against the current in the fast time of 25 days; indeed he made it all the way up to Pittsburgh and thence to Brownsville. By 1816 the Fulton group had gotten three boats up from New Orleans to Louisville and thereby proved that their steamboats could make that diffi-cult trip about as well as Shreve's boats.

In 1816 the much smaller *Despatch* also made it to Louisville, although the Fulton group prevented it from taking cargo. All three of the early Brownsville boats had steam engines designed by Daniel French. However, in 1816, Shreve himself designed a new and much larger boat, the *Washington* (403 tons), which, in spite of an early boiler explosion, became famous as the first great steamboat on the western waters. Shreve captained his new boat up the river from New Orleans to Louisville in 25 days in 1817. Even more striking was the fact, that in the first half of that year, the *Washington*, by making two round trips between Louis-ville and New Orleans, "paid her entire cost and divided $1700 among her owners."[26]

The commercial success of the steamboat led directly to the in-tense competition between many private interests in addition to Fulton and Shreve, and this fierce rivalry invariably produced compromises in safety. The pressure for profits increased the pressure of steam, causing boat cap-tains to operate at higher steam pressures to make greater speed and to avoid delays for maintenance and inspection. These pressures made acci-dents more likely and culminated in the period 1841–1848 with "some

70 marine explosions that killed about 625 persons. In December 1848 the commissioner of patents estimated that in the period 1816–1848 a total of 233 steamboat explosions had occurred in which 2,563 persons had been killed."[27]

These events were directly responsible for changing public opinion concerning the controversial political issue of governmental regulation of private enterprise. Indeed the Constitution of the United States does not expressly sanction by any of its provisions the establishment of federal regulatory agencies. It was strongly held in the early nineteenth century that the government "might act benevolently but never restrictively."[28]

The attitude was changed by the numerous exploding boilers that occurred from 1816 onward, and beginning in 1824 there were repeated attempts to get the Congress to pass restrictive legislation, but each attempt was thwarted or rendered inoperative until midcentury. In 1850, 277 people died in boiler explosions and in 1851 that number rose to 407; clearly Congress had to act and it finally did in 1852.

Three forces converged in that year to get action. Senator John Davis from Massachusetts pushed the new law through Congress, but he could not have done it without the aid of Alfred Guthrie (1805–1882), an engineer from Illinois who, with his own funds, inspected some 200 steamboats in the Mississippi Valley to find causes for boiler explosions. Illinois Senator Shields got Guthrie's report included in the Senate record. The report made recommendations about the same as had an earlier (1846) report from the Franklin Institute (founded in 1824). The *Journal of the Institute* had devoted much space throughout its early years to the scientific study of boiler explosions. Thus did the Senate, the persistent private citizen-engineer, and the Franklin Institute all converge in 1852 to create the joint regulatory agency of the federal government. Guthrie was rewarded for his individual efforts by being appointed the first supervisor of the regulatory agency created through his own efforts.[29]

Ten years after passage of the act, John C. Merriam, editor and proprietor of the *American Engineer,* wrote: "Since the passage of this law, steamboat explosions on the Atlantic have become almost unknown, and have greatly decreased in the west. With competent inspectors, this law is invaluable, and we hope to hail the day when a similar act is passed in every legislature, touching locomotive and stationary boilers."[30]

Merriam's hopes were partly fulfilled in 1887 with the passage into law of the Interstate Commerce Commission, created by Congress to regulate railroads. But the issue of how far government should regulate private industry has remained a live one in U.S. political life.

The owners and masters of steamboats objected strongly to the 1852 law, which slowed construction of new boats and the operation of existing ones. On the other side, insurance companies were organized to insure that steam equipment would be well made and safely used. The reliability of boilers became a major concern of the American Society of Mechanical Engineers, who developed uniform boiler codes.[31] The steamboat had begun the transformation not only of U.S. politics but also of U.S. art.

The Symbolic Perspective

The steamboat was the first powered transportation for passengers and it therefore possessed a unique attraction to large numbers of people. Boat designers quickly realized that a major feature of these machines would be their appearance, both within the cabins and as seen from the shore. Here was the first case of a service for the public where people of no great wealth could for short periods afford to book residence in a moving hotel whose elegance was beyond their permanent standard of living. Thus, these boats quickly became symbols of rising expectations.

At the same time the public began to view the role of government in a new light as a result of the increased boiler explosions and the loss of life. These explosions related to the stresses in the metal boiler wall, which as we have seen depended partly upon the equation $f = Pr/h$ where the higher steam pressure P, the larger boiler sizes (increases in the cylindrical boiler radius r), and the decreases in the boiler wall thickness h all led to a lighter and more economical structure and a more dangerous boiler. Rising expectations competed directly with a more risky environment, both of which the steamboat characterized.

The equation, therefore, symbolizes big government brought into being because the public demanded some control over private enterprises. Soon thereafter the Interstate Commerce Commission would be established as a means to control the railroads, and the steamboat law of 1852 "was used as a precedent to justify regulating legislation in another area where the public interest was threatened."[32]

Not only did the steamboat change government but it also stimulated a new imaginative literature, not now confined to the eastern seaboard with its European influence but rather coming from indigenous American experiences, most especially the steamboat. The first great writer to discover the steamboat and the river as prototypical U.S. symbols was Samuel Clemens (1835–1910), who wrote as Mark Twain. He, too, was deeply affected by the exploding boiler and recounts such horror

Figure 3.8 Mark Twain as a riverboat pilot

anew, not in governmental statistics or scientific formulas, but rather in the personal literary style of a firsthand witness:

> It was six o'clock on a hot summer morning. The Pennsylvania was creeping along, north of Ship Island, about sixty miles below Memphis on a half head of steam, towing a wood-flat which was fast being emptied. . . . There were a good many cabin passengers aboard, and three or four hundred deck passengers—so it was said at this time—and not very many of them were astir. The wood being nearly all out of the flat now, Ealer rang to "come ahead" full steam, and the next moment four of the eight boilers exploded with a thunderous crash, and the whole forward third of the boat was hoisted toward the sky! The main part of

the mass, with the chimneys, dropped upon the boat again, a mountain of riddled and chaotic rubbish—and then, after a little, fire broke out. . . . Many people were flung to considerable distances, and fell in the river. . . . By this time the fire was beginning to threaten. Shrieks and groans filled the air. A great many persons had been scalded, a great many crippled.[33]

This searing account is the subject of a chapter entitled "A Catastrophe" in *Life on the Mississippi*, which Twain published in 1883. The full chapter reports even more horrible events and gives a sense of the tragedies that eventually changed U.S. attitudes politically. Thus we understand transforming events in these three different ways: scientifically as principles in the physics of boilers, socially in the politics of regulation, and symbolically in the art of the imagination.

The western steamboat rose to prominence at the same time that the eastern textile mills dominated U.S. industry, and once again it was the national resource of rivers that made new engineering events possible.

4 Lowell and the American Industrial Revolution

Textiles and Water

U.S. private industry, which began to develop rapidly in the early nineteenth century, depended primarily upon water power derived from small rivers along the East Coast, typified by the Waltham method. This method reached its peak at the new town of Lowell, Massachusetts, founded in 1821 and designed according to a plan developed by Francis Cabot Lowell (1775–1817), who had died four years before. His associates named the new town for him, opened the Merrimack Manufacturing Company for production in 1823, and led the city to become the leading producer of cotton goods (the nation's largest industry) before the Civil War.

Francis Lowell conceived of this new industry during a tour of Great Britain in 1810–1812 when he visited iron works and especially cotton mills in Manchester and Birmingham. His engineering talent allowed him to "smuggle into America valuable mental baggage" of the mechanical basis for those mills. He also returned with a strong desire not to see industry in the United States produce the squalid living conditions and urban blight that were obvious to all visitors to industrial Britain in the early nineteenth century.

The town of Lowell, therefore, got founded on the basis of advanced engineering and social planning. The idea was to have farm girls, strictly supervised, live and work in Lowell, tending the power looms and educating themselves. They were to be only temporary workers, returning to their farm communities with money for a dowry or money to help the family. Lowell was built as an ideal community for both morality and profit. From the 1820s to the 1840s it succeeded in both—not without controversy, but still with impressive profits and a generally good reputation.

Francis Cabot Lowell

Probably no one played a more central role in bringing the industrial revolution to the young United States than Francis Cabot Lowell, seventh generation in Massachusetts, and the one who had done "more perhaps than any other man, to swell the pocketbook of New England and shape its economic future."[1]

Born in the year of Paul Revere's ride and Bunker Hill, Lowell graduated with highest honors in mathematics from Harvard University in 1793, at about the same age as had his grandfather in 1721 and his father in 1760. It was then usual to enter Harvard at age 13. Francis was "very accurate in calculating and projecting eclipses." But he quickly devoted his mathematical talent not to astronomy but to commerce, setting himself up as a merchant in a Boston shipping business.

His grandfather the Reverend John Lowell (1704–1767) had been a well-known Congregational minister in Newbury on the Merrimack River, and his father, John Lowell (1743–1802), known as "the old judge," was a prominent lawyer and member of the 1782 Continental Congress in Philadelphia. Since Percival Lowle (1571–1664) had settled in Newbury in 1639, the family had been prominent around Boston in community affairs, commerce, and religion. Several of Francis's ancestors had written poetry and one was a prosperous shoe manufacturer. "The old judge" was the first noncleric to be appointed to the Harvard Corporation, beginning a family tradition of direct involvement with its school that still exists. The family continued his tradition and developed in addition a literary tradition unparalleled in U.S. history, beginning with James Russell Lowell (1819–1891), who was a renowned poet, a professor at Harvard, and a diplomat, being U.S. ambassador to England among other posts abroad. Abbott Lawrence Lowell (1856–1943) became president of Harvard in 1909, where he served until 1933, and Abbott's sister Amy Lowell (1874–1925) was an important poet in the early twentieth century. Robert Lowell (b. 1917), of the same family, was a major poet during the period following World War II.[2]

By the time of his father's death in 1802, Francis was on solid ground financially, but during the next few years he began to suffer ill health even as his fortune increased. Following the precedent of his elder brother John (1769–1840), Francis decided to take his family to Scotland, where he could recuperate. But he also had another motive.[3]

Over the next two years, Lowell visited cotton mills in Manchester and Birmingham and developed two major ideas about them. He believed, first, that he could duplicate and even improve on their design in New England and, second, that some way had to be found to avoid the bad labor conditions in those industrial cities of Lancashire. Both ideas implied actions of great originality and daring. What would have motivated a man of weak health and comfortable wealth to plan such innovative directions?

Lowell did not write down his motives, or even his plans, but his subsequent actions and contemporaneous events in Scotland help clarify his motives. He memorized the complex machinery used in *carding* (the process of disentangling the cotton fibers after the cotton gin has removed the seeds and hulls), in *spinning* (drawing out and twisting of the fibers into thread or yarn, usually onto a spindle or rod holding a spool), and in *weaving* (interlacing the threads or yarn into cloth). As it was illegal then in Britain to take drawings, machinery, or even machinists out of the country, Lowell had to keep those moving parts in his mathematical mind, which he did. But more remarkably he imagined a new system (not then introduced in Britain) by which all activities from the gin to the cloth would be done in one mill. How, then, could he solve the terrible social problems of factory workers: men, women, and children laboring endlessly for bare subsistence and living crowded together as a "corrupt and debased lower class" in filthy communities that produced beggars, thieves, and drunks?[4]

Thomas Jefferson had likewise worried about such conditions and had concluded that industry should stay in Europe. But Lowell had another solution, and it was suggested by events going on during the early nineteenth century in Scotland. For about 100 years the Scots had been building planned villages by moving people from unproductive farms and squalid hamlets—such as the one in which Thomas Telford was raised—to larger towns newly designed and built by the aristocracy, whose goal was to create clean, modern living conditions and more profitable activities such as fishing and factory industry. There was a dual motive: to raise the living standard of the lower classes and to preserve the social order with the aristocracy securely in charge. The French Revolution had raised fears everywhere of depressed lower classes tearing down the social order, and Lowell's future partner, Nathan Appleton (1786–1862), had mused on this fear in letters home during an 1802 visit to Britain. Appleton had conjectured that the wealth and conveniences of British society were "the consequence of the debasement of the lower classes of society—for the happiness of our country at large I could wish it long without them."[5]

Figure 4.1 Hannah Jackson Lowell (1776–1815)

Lowell, therefore, concluded that an improved and hence profitable textile industry was not enough; there had to be also an improved working society. His motives were private profits and public welfare; he had not the highly idealistic temperament of Robert Owen (1771–1858), whose industrial town of New Lanark, Scotland, was highly publicized during Lowell's stay in that country. Like Lowell, Owen wanted to improve living conditions and raise the moral level of the workers, but unlike Lowell, Owen developed eventually the viewpoint that profits, or capitalism, was a main impediment to the social goals, and he rejected it. His town failed as a social experiment, as did one called New Harmony that he founded in the United States in 1824.[6]

In spite of later failures, the ideas and realities of the planned villages of Scotland provided an example of what could be tried in New England, and when Lowell returned to Boston just before the War of 1812, he had in his mind a major industrial innovation: the complete factory town.

F. C. Lowell.

Figure 4.2 Francis Cabot Lowell (1775–1817)

In Lowell's view such an innovation required a full understanding of how a cotton mill worked physically; of how its labor force should be organized; and of how the enterprise would benefit his family, Massachusetts, and the United States. The order of these requirements was crucial and Lowell was too cautious and practical to mix them up.

Like many major entrepreneurs of the nineteenth century, Lowell realized that the scientific basis of industry was the highest priority, and he, therefore, concentrated immediately on working out machine designs with his talented chief engineer, Paul Moody.[7] It was probably an advantage that they did not have drawings or models of the British machinery. Instead they were forced to rely on Lowell's memory and to re-create the actions for themselves. In that way they could think on their own how best to design the new machinery that Lowell had purchased for mills along the Charles River rapids at Waltham.

Figure 4.3 Textile mills at Lowell, Massachusetts

Nathan Appleton recalled a story indicating the talent of Moody and Lowell's quickness of mind and decision:

> Mr. Shepherd, of Taunton, had a patent for a winding machine, which was considered the best extant. Mr. Lowell was chaffering him about purchasing the right of using them on a large scale, at some reduction from the price named. Mr. Shepherd refused, saying, "you must have them, you cannot do without them, as you know, Mr. Moody." Mr. Moody replied—"I am just thinking that I can spin the cops direct upon the bobbin." "You be hanged," said Mr. Shepherd. "Well, I accept your offer." "No," said Mr. Lowell, "it is too late."[8]

Thanks to Lowell's quick mind and decisive leadership, the Boston Manufacturing Company formed on October 20, 1813; had power looms successfully operating by the fall of 1814; and in 1815 began to pay out regular dividends to the 11 investors.[9] This success depended upon the means of manufacturing, the marketing of cloth, and the system of labor. Lowell with Moody developed the means, which depended upon water power; Appleton provided access to markets through a commercial company he controlled; and Lowell approached the labor problem with his Scottish observation in mind. The solution he found was as novel as it was practical.

Figure 4.4 Lowell House at Harvard University

Since Lowell had decided to employ young, unmarried women for the unskilled labor, he then had to convince his associates not only to provide company-owned boarding houses but also to take on responsibility for the moral lives of these workers. Moreover, these women were, relatively speaking, well paid and paid in cash (rare at that time) and were expected to leave after earning enough money either for their own dowry or for help in their family's finances. The work was not easy, the hours were long, and the controls—by bell—were strict, but it was an attractive bargain and Waltham soon had a waiting list.

Lowell's motives were complicated but his priorities were clear: A profitable business came first but it went together with a labor force that was to be healthy and content, thus avoiding the threat of revolution that he had felt growing in Britain during his two-year stay. Profit also required recognition of competition, and Lowell realized that his new industry could not compete with cheap imports from India. Following the end of the War of 1812 British cotton goods once again flooded the U.S. market, putting many New England textile companies in danger.

Lowell travelled to Washington, D.C., in 1816 to convince the Congress to put on a tariff. He wanted it high enough to exclude cloth from India but he believed that his less fancy cloth could compete with the British goods without such a high tariff as to exclude them. In this compromise way, he could ensure manufacturing profits without alienating the commercial traders of Boston. By diplomatic skill he succeeded in convincing powerful southern politicians such as John C. Calhoun, and the Congress passed its "infant industries" tariff legislation. By 1817 the modest tariff, the efficient production, and the vigorous sales combined to make Waltham highly successful, paying a dividend of 20 percent. Lowell had succeeded in founding a major industry, but at the price of his health. It broke that same year and on August 10, 1817, he died.[10]

The First Industrial City

The immense success at Waltham stimulated expansion there, but soon the surviving partners realized that the potential water power of the Charles River would not be enough to allow more growth. Another site was needed if the Boston Associates were to establish a major new industry in New England. That new site had to satisfy at least three main criteria: potential water power; transportation to markets, especially Boston; and availability of land for building a large complex.

In November 1821, Nathan Appleton, Patrick Jackson (Lowell's brother-in-law), Paul Moody, and three others visited the farming region around the Pawtucket Falls of the Merrimack River some 28 miles northwest of Boston.[11] The river was suitable for water power because of its relatively large flow, measured in cu. ft. of water per min. and later determined to be, at low water, 216,000 cu. ft. per min. The visitors could see the large flow even if they could not then accurately measure it. They could also see the substantial drop in the level of river as it made a 90-degree bend at the little village of Chelmsford, where the Concord River joined the Merrimack (see page 75). That drop, called the head, was 33 ft., which contributed to the power potential. This may be defined as the prod-

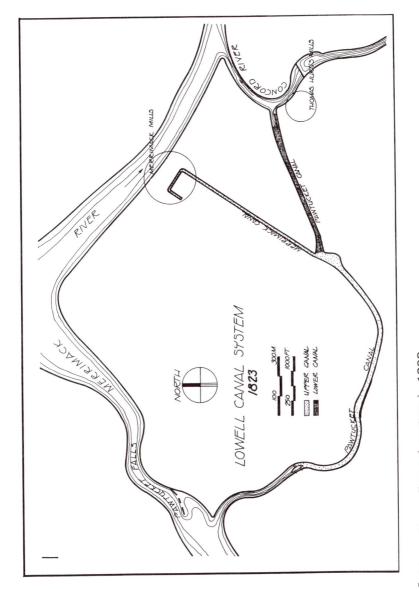

Figure 4.5 Map of Lowell canal system in 1823

uct of flow, head, and the density of water, 62.4 lbs. per cu. ft. Thus the power potential at the site was approximately

$$216,000 \times 33 \times 62.4 = 445,000,000 \text{ ft.-lbs. per min.}$$

Using Watt's definition of 1 Hp equaling 33,000 ft.-lbs. per min., we can estimate the Pawtucket Falls power potential to be

$$\text{Hp} = \frac{QwH}{33,000} = \frac{445,000,000}{33,000} = 13,500 \text{ Hp}$$

with Q the flow, w the density of water, and H the drop or head. As late as 1855, the total power derived from water in all of France was only 20,000 Hp. Surely this site in Massachusetts was ideal for a power source.[12]

In addition to power, a means of transportation, the Middlesex Canal, stood ready and greatly underused, linking the Merrimack River just above the Pawtucket Falls to Boston 28 miles away. The canal had been a financial failure since its completion in 1803, but it could now serve to bring cotton to the new mills and to return cloth to the market center at Boston.[13]

Finally, the land in the region was in no demand by 1821, and so long as the associates operated secretly in buying it up, the price would be low. They did, and by 1823 the first large-scale mill owned by the Merrimack Company began to produce cloth; by 1825 the new company was paying dividends.[14]

In 1822 Appleton, Jackson, and their associates bought control of the proprietors of Locks and Canals, a corporation chartered in 1792 that had constructed the Pawtucket Canal bypassing the Pawtucket Falls and including four locks. It was a failure because the Middlesex Canal, completed in 1803, took most of the trade to Boston. The proprietors were happy to sell out in 1822. Thus, having full control of the region, the association, through the new Locks and Canals Company, built a temporary dam above Pawtucket Falls, widened the Pawtucket Canal, and built the new Merrimack Canal, the first power canal to bring water to the Merrimack mills 30 ft. above the river level there.

By 1825 the associates realized the need for expansion. They estimated that there could be 50 times the power then used in their first mill, so the Locks and Canal Company embarked upon an ambitious plan to create a network of power canals unlike anything ever seen.[15] Meanwhile Moody, having moved from Waltham in 1822, established a machine shop at Lowell and began to make all the textile machinery for the second and third mills of the Merrimack Company.[16] The associates reformed the Locks and Canal Company by selling them the machine shop, the canals,

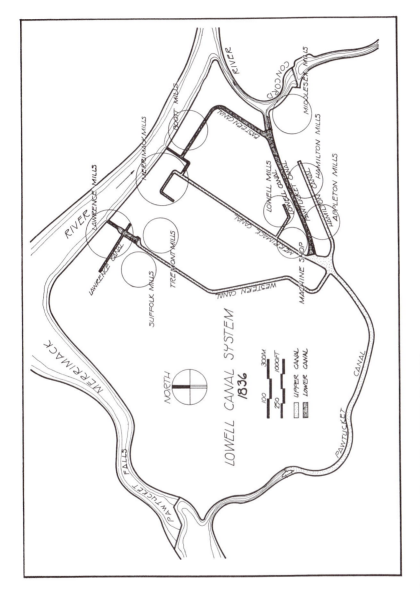

Figure 4.6 Map of Lowell canal system in 1836

and much land, thus separating them from the Merrimack Company, which could now focus solely on textile production.

Between 1825 and 1835 the proprietors completed their grand scheme, by which time 10 separate cotton-mill companies were operating mills on a network of canals either with a single fall of 30 ft. or with falls of 13 ft. and then 17 ft. At this same time the associates, under Jackson's leadership, built a railroad from Boston to Lowell, opened in 1835, to provide better transportation to and from the markets. For this they imported in 1832 two steam locomotives from Robert Stephenson, and once the locomotives arrived they studied all parts so the Lowell machine shop could then build their own. For that they hired George Washington Whistler (1800–1849), already by 1834 a veteran railway engineer, to build engines in the machine shop. His son, the painter James A. M. Whistler, was born in Lowell that year. When the railroad opened in 1835, a Whistler-built engine entitled *Patrick* made the Boston run. It could not be named *Jackson* for fear of confusion with the president of the United States, for whom the associates had no sympathy at all.[17]

When Whistler resigned in 1837, eventually to go to Russia to build railways there, his young assistant, James B. Francis (1815–1892), was designated as engineer for the proprietors and the nature of engineering at Lowell underwent a radical change. The primary reason for change came because of increased competition and the resulting pressure for higher productivity. Up to 1837 the power system was based upon ancient engineering ideas: canals, gates, and waterwheels. More significantly, little attention was paid to predicting water supply and determining water flows accurately. Measurements of waterwheel efficiency were, therefore, uncertain. Between 1837 and 1855 Francis would transform all that by introducing into hydraulics a degree of precise testing and careful design that would put a major branch of engineering on a firm scientific basis.

5 Francis and the Industrial Power Network

Francis Arrives

James Bicheno Francis, born in Southleigh, Oxfordshire, England, on May 18, 1815, began working at age 14 with his father on railways and canals. Like many young Englishmen of the time, Francis decided to come to the United States, where much larger civil works projects were beginning. In 1833 he arrived at New York and applied to George Washington Whistler, then one of the most prominent engineers in the country. Whistler took him on as an assistant in the construction of the New York, Providence, and Boston Railroad. When the Boston associates invited Whistler to Lowell in 1834, he took Francis with him to work for the Locks and Canals Company. Francis made such a strong impression on the management that when Whistler left in 1837, the 22-year-old Englishman was made engineer for the company.[1]

The main network of power canals was by then serving the 10 cotton-mill companies, but between 1837 and 1845 there arose the need to make the system more efficient. The mills all had large waterwheels that converted the potential power to mechanical power at an efficiency of about 60 percent.[2] In addition, the amount of water delivered to each mill was in dispute. A textile company would claim, if their power dropped, that it was not getting its share of the water. Also because of competition, the companies pressed for greater productivity by installing more machines and running them faster, all of which required more power.[3]

To help settle disputes about water flow, the proprietors in 1841 hired James F. Baldwin, Whistler, and Charles S. Storrow, each an eminent engineer, to determine the flow of water taken by each company. In their final report of December 17, 1842, these engineers recommended

Figure 5.1   James B. Francis (1815–1892)

that the flow, called Q, be determined by measuring the velocity of the water on the surface, called V, and multiplying it by the cross-sectional area, called A, of the water in the canal (width B times water depth D: $A = BD$) and then reducing the resulting product by a constant, C. They determined this constant empirically for each canal.

Francis had been engineer for the proprietors during these tests and he would later publish the results.[4] Over the next decade Francis raised serious doubts as to the validity of those results, especially because there had been no convincing way to establish the ratio between the mean velocity of the water and the velocity at the surface. Indeed the three engineers' value of C was their estimate of that ratio, but it did not satisfy Francis.

Another worry was the unusual periods of dry weather in the early 1840s, which dropped the flow of the river and hence the potential power from the canals. In 1845 the Locks and Canals Company bought water rights to control the flow from lakes in New Hampshire that fed the Merrimack and thus help conserve water for use in those dry seasons.[5]

Issues of supply, measurement, and efficiency caused the proprietors of Locks and Canals in 1845 to reorganize their company and give control back to the textile corporations using the canals. They elevated Francis to agent as well as engineer for the new company, for which he thus became not only chief engineer and designer but also chief arbiter of water rights. He was later referred to as "the chief of police of water."[6]

Francis Rebuilds Lowell Power

Once confirmed as the sole director for all water power, Francis undertook the most ambitious plan for industrial research and development in the United States up to that time. He redesigned the power canal system, laying out a new and far larger canal, the Northern Canal, to relieve the old Pawtucket Canal by providing a much greater area of canal section and hence room for a greater flow. For example, Kirk Boott had rebuilt the Pawtucket Canal to be 60 ft. wide, 8 ft. deep and carry a flow of about $Q = 2,500$ cu. ft./sec. The Northern Canal, by contrast, is 100 ft. wide and 15 ft. deep. By 1853, when Francis had restructured the system, it provided a minimum of 11,900 Hp potential, equivalent approximately to a flow of $Q = 3,500$ cu. ft./sec., easily carried by the new canal on a fall of $H = 30$ ft.[7] The power potential is

$$\text{Hp} = \frac{QwH}{550} = \frac{3,500 \times 62.4 \times 30}{550} = 11,900 \text{ Hp}$$

Because Q is in cu. ft./sec., the denominator must be converted to ft.-lbs./sec. by dividing 33,000 by 60.

The Northern Canal was capable of carrying a much larger flow on the basis of its much greater cross-sectional area. At less than half full, it could carry sufficient flow to develop the full potential water power of 11,900 Hp.[8]

But the construction of this canal required new head gates and the rebuilding of a section of the river dam, all to divert a controlled amount of river water into the Northern Canal. Francis designed a special water turbine to control the gates, and he also installed large testing chambers and other equipment that he planned to use for experiments. Francis took the opportunity to create a new kind of laboratory for full-scale testing.[9]

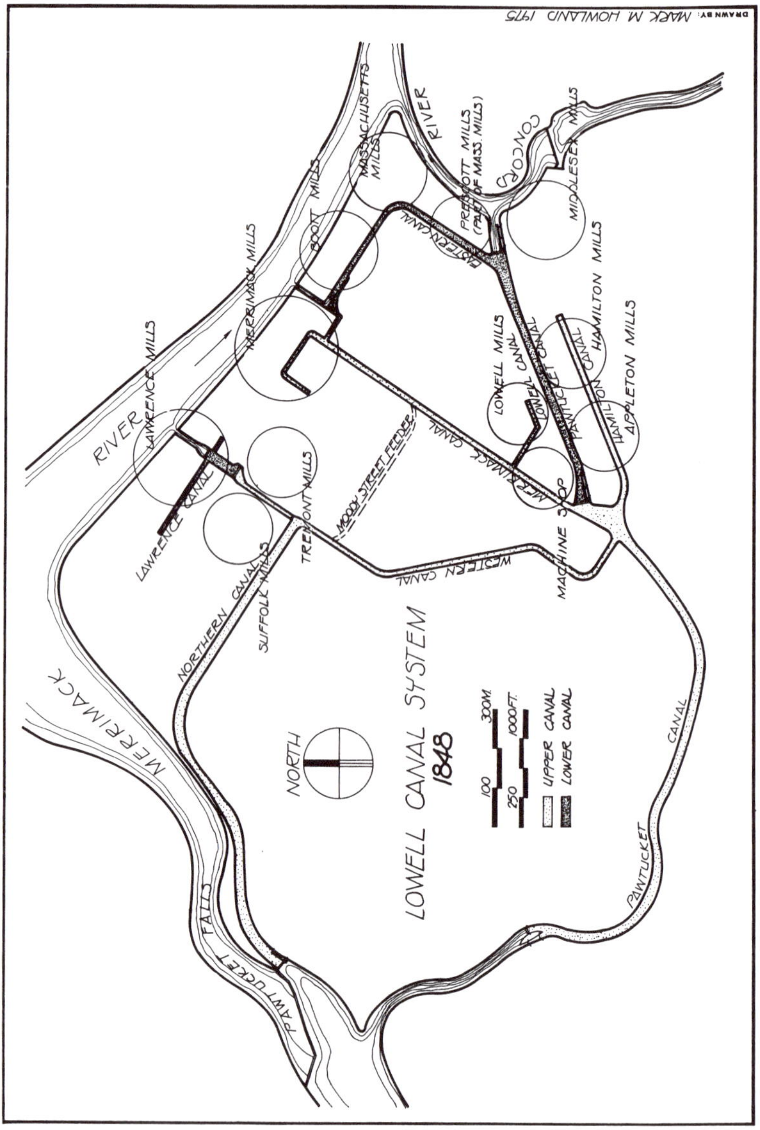

Figure 5.2 Map of Lowell canal system in 1848

But Francis also imagined the new canal to be a park as well as a utility and a laboratory. He lined the banks of the canal with trees and it became a favorite site for walks. One resident described the scene:

> Years ago, it was one of the sights of the city, for its citizens on Sunday afternoons to promenade from Cabot Street along the side of the Northern Canal, up toward Pawtucket Street under the high bridge; and along the stone dividing wall, between the canal and the river, thence to the Pawtucket Falls Bridge. The writer clearly recalls the hundreds, who enjoyed the walk, which presented a singularly attractive view of its entire length, and was known as the Canal Walk.[10]

Boyden, Francis, and the Turbine

Francis had improved the water supply Q substantially with the completion in 1848 of the Northern Canal, but the problem of efficiency was different. It involved replacing the old wooden waterwheels by a water motor designed to increase the power to the textile machines. In 1844 the Appleton Mills purchased a turbine designed by Uriah Boyden (1804–1879) that clearly increased efficiency, but by how much?

The problem was to make wheels to use energy efficiently, and Boyden's 1844 turbine for the Appleton Company produced 75 Hp with an efficiency of 78 percent, according to Boyden. Then in 1846 he designed three more turbines for the same company. Here he made a contract for his payment as designer to be in the amount of $1,200 plus $400 for each percent of efficiency that the turbines tested above 78 percent. Boyden designed the tests and Francis made the calculations and certified that the turbines reached 88 percent so that without hesitation the company paid Boyden $5,200.[11] Clearly the increased efficiency was worth money to the designer and much more in profits to the owner.

Following that success, Boyden designed a large number of such turbines throughout New England, and in 1849 at Francis's urging the manufacturing companies at Lowell bought the rights to Boyden's turbines and Francis proceeded to design new ones, direct their construction, and plan tests in his newly built laboratory.[12] The first such tests in February 1851 were on one turbine for the Tremont Mills that had been completed in January 1851. The water power available was $P_{in} = QwH$, for which, in test number 1, Francis calculated 111,870 ft.-lbs./sec.[13]

Francis used a Prony dynamometer, invented in France and improved by Boyden, to measure the mechanical power delivered by the turbine. Francis clamped this Prony brake onto the vertical shaft turned by

Figure 5.3 Uriah Atherton Boyden (1804–1879)

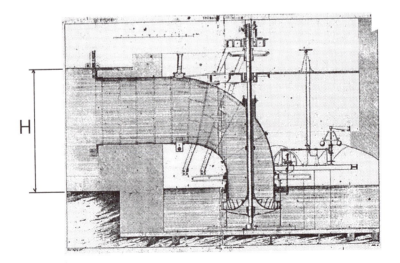

Figure 5.4 Cutaway view of the Tremont turbine

Turbine Tests by Francis

Power in (P_{in}) depends upon the *flow* of water (Q), its *density* (w), and the *head* (H) or drop in water height:

$P_{in} = QwH$

Flow (Q) is the volume of water flowing past one point in 1 sec. In this case Francis (using a weir) measured the flow to be 139.42 cu. ft. per sec.

Density (w) of water is 62.4 lbs./cu. ft. (Francis used 62.375).

Head (H) is the difference between the height before it enters the turbine and the level where it goes out. At the Tremont Mills $H = 12.864$ ft.

$P_{in} = 139.42(62.375)12.864 = \textbf{111,870 ft.-lbs./sec.}$

Power out (P_{out}) depends upon the product of *thrust* (T) and the turbine *velocity* (V):

$P_{out} = TV$

Thrust (T) Francis measure to be 1,443.34 lbs. at the outside of the wheel.

Velocity (V) Francis found when the Prony brake slows the turbine velocity down to 0.894 revolutions per sec.

The brake wheel radius was 10.83 ft. so that the circumferential distnace in one revolution was $2\pi r = 68$ ft. Therefore, the velocity of the outside of the wheel was

$V = 0.894(68) = 60.8$ ft./sec.

and hence the mechanical power delivered by the turbine shaft (power out) was

$P_{out} = TV = 1,443.34(60.8) = \textbf{87,760 ft.-lbs./sec.}$

Efficiency was

$$\frac{P_{out}}{P_{in}} = \frac{87,760}{111,870} = \textbf{78.4\%}$$

Measuring Turbine Power Out

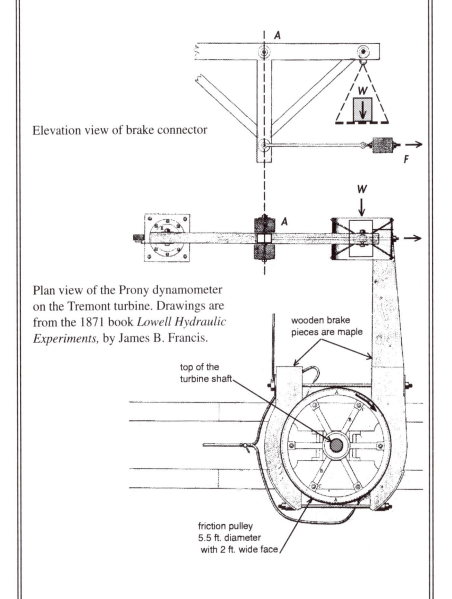

Elevation view of brake connector

Plan view of the Prony dynamometer on the Tremont turbine. Drawings are from the 1871 book *Lowell Hydraulic Experiments,* by James B. Francis.

wooden brake pieces are maple

top of the turbine shaft

friction pulley 5.5 ft. diameter with 2 ft. wide face

Model for the Prony Dynamometer

(1) Elevation

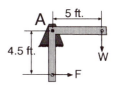

Rotation about point A

$5W = 4.5F$

$$F = \frac{5}{4.5}W = 1.11W$$

(2) Plan

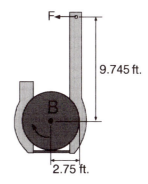

Rotation about point B

$2.75T = 9.745F$

$\quad\quad\ = 9.745(1.11)W$

$\quad\quad\ = 10.83W$

$$T = \frac{10.83}{2.75}W$$

(3) Turbine power

$P_{out} = TV$ (ft.-lbs./sec.)

$T =$ (lbs.) – traction (friction) between the pulley and the brake

$V =$ rev./sec. $\times$ ft./rev. – velocity of the friction wheel

$V =$ (rev./sec.) $\times (2\pi \times 2.75)$

$$T = \frac{10.83}{2.75}W$$

$T = \Sigma t$

$P_{out} = (10.83 \times 2\pi)W \times$ (rev./sec.)

$\quad\quad\ = 68W$ (rev./sec.)

Francis measured the revolutions per sec. of the turbine shaft and the corresponding weight W from which he then could compute the power produced from the turbine.

LOWELL

HYDRAULIC EXPERIMENTS.

BEING A SELECTION FROM

EXPERIMENTS ON HYDRAULIC MOTORS,

ON THE

FLOW OF WATER OVER WEIRS, IN OPEN CANALS OF UNIFORM RECTANGULAR
SECTION, AND THROUGH SUBMERGED ORIFICES AND DIVERGING TUBES.

MADE AT LOWELL, MASSACHUSETTS.

BY

JAMES B. FRANCIS,

CIVIL ENGINEER, MEMBER OF THE AMERICAN SOCIETY OF CIVIL ENGINEERS AND ARCHITECTS,
FELLOW OF THE AMERICAN ACADEMY OF ARTS AND SCIENCES, MEMBER
OF THE AMERICAN PHILOSOPHICAL SOCIETY, ETC.

THIRD EDITION.

REVISED AND ENLARGED, WITH MANY NEW EXPERIMENTS,

And Illustrated

WITH TWENTY-THREE COPPER-PLATE ENGRAVINGS.

NEW YORK:

D. VAN NOSTRAND, PUBLISHER, 23 MURRAY STREET AND 27 WARREN STREET.

1871.

Figure 5.5 Title page from *Lowell Hydraulic Experiments*

the turbine. As the brake pressed against the wheel it developed a friction force T, and that slowed the velocity of the wheel V. The product of T and V also gives power just as it does for the paddles of the steamboat. In this first test Francis calculated the power from the turbine, $P_{out} = 87{,}760$ ft.-lbs./sec.[14]

The efficiency, therefore, was

$$\frac{P_{out}}{P_{in}} = \frac{87{,}760}{111{,}870} = 0.784$$

or 78.4 percent. Francis ran 92 such experiments on the Tremont turbine in which he varied, among other things, the friction force on the brake, which in turn changed the velocity; hence for each new case he got different values of mechanical power.

For turbine design it is essential to know how efficiency varies with the turbine velocity, and Francis plotted such results between the limits of $T = 0$ (when the velocity of the unloaded turbine was a maximum of 121.5 ft./sec.) and $V = 0$ (when the load $T = 4{,}080$ lbs. was just large enough to stop the shaft from turning).[15] Clearly at these two limits the mechanical power is zero, and Francis's curves showed that the turbine reached maximum efficiency at velocities about half the maximum, or

$$\frac{P_{out}}{P_{in}} = 0.794 \text{ (test \#30)}^{[16]}$$

From these detailed and careful experiments Francis derived "Rules for Proportioning Turbines" that, as he put it, "should only be taken as guides in practical applications until some more satisfactory are proposed, or the intricacies of the turbine have been more fully unravelled."[17] He knew his rules had a sound basis in his experience and his tests, but he also recognized that they were limited. He knew his results were not general, but by being fully reliable they have endured. Indeed, with the coming of hydroelectric power in the late 1880s, engineers rediscovered the work of Francis and many of the major turbine installations during the twentieth century used what came to be called Francis turbines.[18]

Profits, Equity, and the Weir

Francis's calculations of turbine efficiency depended not only on his "rules" but also on his ability to measure the flow into the water motor. Textile profits closely corresponded to the productivity of power machinery, and the companies wanted to know how much potential power they were getting and to pay for no more. Measurements of Q, the flow, became critical to Francis's role as agent for the companies and policeman for water equity.

The problem was to determine Q reliably without disrupting the flow to the mills. Earlier efforts were suspect because they depended upon measurements of water velocity V, because $Q = AV$ where A is the cross-sectional area of the water in the canal. Francis needed a formula that depended not upon velocity but upon a simple measurement of the water height; after exhaustive tests he developed his weir expression in terms of H, the vertical distance between the crest and the undisturbed water level (see the figure).

In 1852 Francis directed a series of tests in the lower locks at Lowell in which he carefully measured the flow over a weir; the height of the undisturbed water above the weir crest, H; and the length of the weir (in the direction of the canal width), L, where, for ideal flow,

$$Q = AV = (HL)V$$

where A is the area of the cross-section of the weir through which the water flows and V is the mean velocity of flow in ft./sec. This meant that hydraulic theory, well known to Francis in 1852, gave $V = (\frac{2}{3}) \sqrt{2gH}$ so that for ideal flow

$$Q = \frac{2}{3} \sqrt{2gH} \, (HL) = 5.35LH^{3/2}$$

where $g = 32.2$ ft./sec.^{2}, the acceleration of gravity, and H is the height of water in the channel measured from the crest of the weir (see figure).[19] The actual flow will never be as great as the ideal flow predicted by theory because of the loss in energy as the flow contracts over the crest, where its height is always less than H. Francis knew this well as did many earlier engineers; what he set out to do was to determine a reliable factor, C (to replace the theoretical value of 5.35), that company officials, for whom Q had a direct financial meaning, could rely upon.

To see better how Francis's mind worked and in general how engineers think, we need to include one more term in the equation of flow and describe how Francis disposed of it. Usually a weir has vertical sides that constrict the flow of the channel, meaning the weir length L is less than the channel width, and the effect is to disrupt the smooth flow and hence reduce its quantity (see figure). Boyden had suggested to Francis in 1846 that the reduction be expressed by $L' = L - bnH$ and that the 3/2 exponent of H needed to be found empirically. Thus the quantity of flow was

$$Q = C \, (L - bnH) \, H^a$$

Model for Measurements of Flow

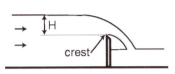 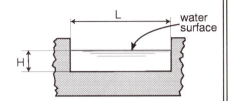

The *flow*, Q, depends upon the cross-section area, $A = LH$, and mean velocity, V.

$$Q = AV = (LH)V$$

For *ideal flow*, $V = \frac{2}{3}\sqrt{2gh}$, where $g = 32.2$ ft./sec.^{2}, the acceleration of gravity.

$$Q = \frac{2}{3}\sqrt{2g}\,LH^{3/2} = 5.35LH^{3/2}$$

For *actual flow*, Francis found from tests that 5.35 should be taken as 3.33 and $L' = L - 0.1nH$.

$$Q = 3.33(L - 0.1nH)H^{3/2}$$

This Q can be found solely from the dimensions of the weir and the height, H, of the undisturbed water level behind the crest of the weir.

Weir dimensions for test #1

$L = 9.997$ ft.

$n = 2$

Water level above weir crest

$H = 1.533$

Measured flow

$Q = 61.3$ cu. ft./sec.

Calculated constant

$$Q = C[9.997 - 0.1(2)1.533]\,(1.533)^{3/2}$$
$$61.3 = C(9.69)(1.9)$$
$$C = 3.33$$

which was the value that Francis proposed for the general formula of flow through a rectangular weir.

Weir Studies of James Francis

Measurements of Flow

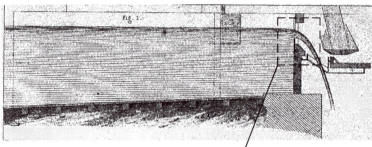

Section view of one of several weirs
in the experiments at the lower locks.
(Drawing is from the 1871 book
Lowell Hydraulic Experiments,
by James B. Francis.)

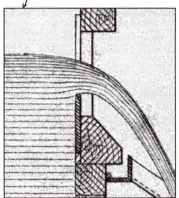

Carefully detailed drawing
shows how the level of water at
the weir drops lower before the
gate and flow tapers beyond
the gate.

where a and b are constants to be found from tests and n is the number of vertical sides to the weir.[20]

First Francis pointed out that this formula makes no sense if $bnH = L$ because then $Q = 0$. Thus, this simplified empirical formula "cannot be of general application" and it should be used, Francis recommended, only where L is not less than $3H$. "In practical applications, this will seldom be an inconvenience."[21] The problem was to find a and b.

Second, Francis assumed a to be unknown, following Boyden, instead of $\frac{3}{2}$, or 1.5, as derived from theory. This suspicion of the theory led Francis to carry out these first experiments in early 1851, which revealed a to be 1.47, or very close to the theory.[22] When Francis made his second set of tests in late 1852 he simply assumed $a = 1.5$, for as he said, "the use of a fractional power, such as $a = 1.47$ is very inconvenient and, to persons not well skilled in the use of logarithms, offers great difficulty."[23] Moreover, "after many trials of other values" he adopted $b = 0.1$ as an accurate number for interpreting the results from the second set of tests.[24]

Now Francis could compute that empirical coefficient C using careful experiments with known quantities of water flow Q, height H, and length L. From 88 separate tests, Francis found values of C from 3.30 to 3.36.[25] He concluded that $C = 3.33$ was reasonable; it was also the arithmetical mean of all 88 results.

Francis concluded his report on these tests by comparing his results to those made in France between 1827 and 1846 to show that his formula with $C = 3.33$ and $b = 0.1$ could be applied there as well.[26] He also included a cautionary note defining clearly the limitations of his results.[27] His thorough work established a standard that is still in use today.[28] His goal was not only to satisfy business contracts to manufacturers but also to share his results with the profession.

Francis was a modern engineer, first because his design for weirs and turbines came from calculations based upon scientific study of full-scale performance. Second, the calculations themselves are correct by today's standards even though new experimental data now exists and more sophisticated procedures are available to solve equations. Third, Francis focused on full-scale design under the continual stimulus of business economics. He was not interested primarily in new knowledge but rather in new designs that would profit his employer. He was recognized as being perhaps the foremost hydraulic engineer of his day, confirmed by his election in 1880 as president of the American Society of Civil Engineers (ASCE).

Theory and Practice for Francis

One anecdote illustrates Francis's mind, with its suspicion of theory and yet its insistence on precise scientific experimentation. It has nothing to do with hydraulics but rather with a mild controversy in the early nineteenth century over the proper value for the deflection of a beam clamped at either end and loaded at midspan by a concentrated weight (*P*). Peter Barlow (1776–1863), the British engineer, had written in 1833 that the midspan deflection of such a case was two-thirds of the value gotten for a simply supported beam (ends free to rotate at supports). On the other hand, Navier (1785–1836), the famous French theoretician, had determined the ratio to be one-fourth not two-thirds, and in 1872 Francis presented a paper at the fourth annual convention of the ASCE in which he took up the discrepancy. "Not feeling competent to decide which, if either, of these eminent authorities was correct, and having occasion to apply it in practice," he devised some careful experiments by which he showed results that "agreed substantially with Navier."[29] Here Francis ventured into a new field and, unwilling to accept the by-then fully developed theory of beams, he confirmed a well-known result by testing. His British heritage perhaps tended to esteem Barlow with Navier even though Barlow's works had been discredited.[30]

Still, as a practitioner, such an approach by Francis was typical of the best nineteenth-century U.S. engineers in their willingness to look at works from abroad and then determine for themselves what was a useful result. It may have led to some unnecessary testing, but its fruit was reliable and innovative design that often went beyond the achievements of Europeans. Another example of such achievements came with the railroad, the new U.S. transportation system that had attracted Francis to the United States in the first place.

II

Crossing the Continent, 1830–1883

6 The Stephensons, Thomson, and the Eastern Railroads

From Water to Steam

The transformation of water into steam was the leading change in mid-nineteenth-century U.S. industry. Water power, which dominated Lowell and the rest of the United States before the Civil War, gradually gave way to steam power such that by 1869 more than half of all industrial power was by steam. In 1840 nearly all such power was by water.[1]

In a similar manner, water dominated transportation within the young United States, beginning with Fulton's 1807 franchise to Albany, continuing with the vigorous growth of Mississippi commerce up to mid-century, and including along the way the Erie Canal of 1825. And just as the steam engine began to transform industry by the 1850s, so did it change transportation by means of the central cultural event of that century, the railroad.

Just as with industrial machinery, the locomotive and the building of rail lines originated in Great Britain, and the new American republic imported both the artifacts and the ideas as rapidly as it could starting in 1830, at the same time that the first major line opened in Great Britain.

British Beginnings: The Road to Rainhill

The first steam locomotive was run in 1804 at the Pendaron Iron Works in South Wales. Designed by Richard Trevithick (1771–1833), it had a high-pressure (probably between 30 and 50 psi) steam boiler and a cylinder 8.25 in. in diameter with a stroke of 4.5 feet. Trevithick designed a similar locomotive for a coal installation near Newcastle in 1805 but he

Figure 6.1   George Stephenson (1781–1848)

never succeeded in making a commercial rail line for passengers or general freight.

The true pioneer in railroad design was George Stephenson (1781–1848) from the north of England. He built his first locomotive in 1814. In 1821 a charter was granted for a horse-drawn rail line between Stockton and Darlington, 30 miles south of Newcastle. Stephenson was hired as a civil engineer to lay out the line in 1821 and he decided to try out his locomotive. In 1823 the charter was changed to allow steam-powered locomotives, and in 1824 Stephenson ordered that two be built. The line opened in 1825 and demonstrated quickly the potential for this new mode of transportation.

Stephenson's little 12-mile line used horses frequently and was thus not immediately hailed as a great event by the public; that would occur five years later when George and his son Robert (1803–1859) completed the Liverpool & Manchester Railway. This major line was built over the objections of the turnpike companies, of the "horse-loving country gentry," and of the canal interests (especially those connected with the duke of Bridgewater's midcentury canal).

In 1826 George Stephenson had been asked to design the railroad between Liverpool and Manchester, two major English cities with great industrial wealth. There arose a conflict on the board of directors over the mode of locomotion. Some wanted steam locomotives only whereas others argued for cables to pull the trains by stationary steam engines up the inclines on the route. There was still doubt in many people's

Figure 6.2 Robert Stephenson (1803–1859)

minds that smooth iron wheels riding on smooth iron rails could actually exert enough traction to pull heavy engines and cars on straight runs, let alone up inclines.

Because of this doubt, the directors organized a competition to be held on the Manchester side of the Rainhill Bridge about 10 miles from Liverpool, in order to prove that steam locomotives could carry heavy loads and pull them over long distances. The requirements were that the locomotives carry 20 tons of load at a minimum speed of 10 mph over the course 40 times, for a distance of 60 miles, which was that between Liverpool and Manchester.

The company expected many entries, but only three locomotives competed. Stephenson entered a new locomotive, the *Rocket*, designed and built by Robert Stephenson and Co. The *Rocket* easily won, reaching speeds of nearly 30 mph near the end of the trial.[2] One of the other locomotives was designed by Braithwaite and Ericson; John Ericson was later

Figure 6.3 Rocket engine of Robert Stephenson

to gain fame as the designer of the ironclad boat the *Monitor*, which fought in the U.S. Civil War.

Parliament had ridiculed George Stephenson for his "absurd claims" and his "country ways and speech," but his persistence eventually won him a charter from the London politicians for the Liverpool and Manchester line. He constructed an impressive line by bridges, through deep cuts, and across the "dreaded and supposedly bottomless bog called Chat Moss."[3]

The Liverpool and Manchester Railway opened on September 15, 1830, with George Stephenson driving the lead locomotive, the "*Northhumbrian*."[4] The gauge, or distance between the insides of the rails, was made 4 ft. 8.5 in. on this first major railroad. The origin of this strange dimension is to be found in the old wooden tramways used in the Northumberland collieries before steam locomotives. At Killingworth near Newcastle, where George Stephenson had made his first steam locomotive, the distance between insides of the rails had been 4 ft. 8 in. Somehow a half inch had been added to the usual gauge by the time George Stephenson laid out the Liverpool and Manchester line and thus 4 ft. 8.5 in. became the standard gauge for rail lines designed by the Stephensons north of London. The same was later adopted for the London and Birmingham line and hence for the main route between London and Liverpool.[5]

Early Locomotive Design

On July 1, 1849, W. P. Marshall wrote a detailed paper, under the supervision of Robert Stephenson, on the mature Stephenson locomotive as perfected during the 20 years that followed the famous success at Rainhill.[6] Marshall noted at the outset that the engine cost was considerable (about 1,400 lbs.) and "the expense is nearly the same as that of doing the same work by horses, if saving of time from the greater speed and heavier loads . . . are not taken into account." But of course the whole point of early nineteenth-century development in society was to get greater speed with greater load-carrying capacity. Marshall described Stephenson's design by emphasizing the difference between mobile steam engines and the stationary ones developed by Boulton and Watt.

Both steam engines developed power by reciprocating motion; again, recalling Watt's definition,

$$Hp = \frac{PLAN}{33,000}$$

For the locomotive it was essential to reduce weight and increase power. To achieve the essential lightness, locomotives needed engines of small dimensions (small L and A) capable of great power (large P and N), which demanded means of producing a large quantity of steam quickly. The structure of the locomotive needed to be "very strongly and firmly made and framed together" so it could resist the high internal forces and the "violent shocks" caused by the speed of such a heavy machine.[7]

Just as with Watt and Fulton, Robert Stephenson created a practical new machine by taking what was available and putting things together in new ways. He sat down, as did Fulton, amid his "levers, screws, wedges, wheels, etc. like a poet among letters of the alphabet," and out of that effort came the Stephenson locomotives.

Marshall explained this synthetic process by stressing that

> The great perfection of the present locomotives, and their superiority to the old ones, is caused not so much by the application of new inventions to them, as by the combination of many former ones, and the uniting together several plans which, separately, would be but of small value.[8]

Stephenson could not have gotten so much steam so efficiently without using numerous small-diameter tubes to carry hot gases through the boiler. In turn these tubes would not have worked well without the powerful chimney draft produced by the blast pipe. All things had to fit together to make this new machine fast, powerful, and safe.

Stephenson's use of hot gas tubes within the boiler gave far greater heating surfaces and hence produced much more steam per pound of coal burned. Stephenson was the first to use heating tubes within the boiler in Great Britain although the idea came initially from France. M. Sequin first took out a patent on the idea in 1828, but Stephenson's use

Stephenson Locomotive of 1835

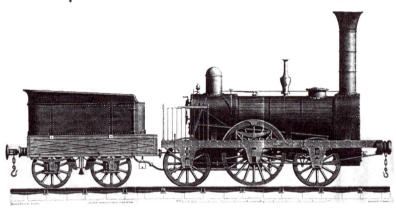

This Stephenson locomotive has a single pair of 5-ft. diameter drive wheels. Each wheel is driven by a steam pressure of 47.5 psi from a 12-in. diameter cylinder with an 18-in. double-acting stroke.

Engine Power

The engine power (P_E) from each of the two driving cylinders depends upon pressure of steam (P), length of stroke (L), area of cylinder (A), and number of strokes per min. (N).

steam pressure $P = 47.5$ psi, double acting

power stroke $L = 1.5$ ft., each power stroke

piston area $A = 113$ sq. in.

speed $N = 157$ power strokes per min.

$$P_E = \frac{\text{PLAN}}{33,000} = \frac{47.5(1.5)113(157)}{33,000} = 38.3 \text{ Hp/cylinder}$$

for two cylinders, $P_E = 2(38.3) = \textbf{77 Hp}$

was not derived from that patent; rather it apparently came as an independent invention.[9]

The tubes carry the hot air into the chimney or smokestack, although in the Stephenson engine described by Marshall in 1849 no smoke was produced (except at first) because the fuel was coke not coal. The

Traction Power

Traction power of the driving wheels acting on rails (of the same form as the steamboat paddlewheels acting on water) depends upon:

Traction force (T), which is related to the friction between wheels and rails, to drag in air, and to track incline.

Velocity (V) of the driving wheel in ft. per min. will depend upon the wheel diameter (D_W) (in this case 5 ft.) and the number of power strokes (N) per min. (314 for the two cylinders combined). The number of power strokes per revolution (rev.) for two double-acting cylinders is 4:

$$\pi D_W = \pi\, 5.0 = 15.7 \text{ ft./rev.}$$

$$\frac{N}{4} = \frac{314 \text{ strokes/min.}}{4 \text{ strokes/rev.}} = 78.5 \text{ rev./min.}$$

$$V = \frac{\pi D_W N}{4} = 15.7(78.5) = \mathbf{1{,}230 \text{ ft./min.}}$$

which converts to mph (miles per hour):

$$V = 1{,}230\, \frac{60 \text{ min.}}{5{,}280 \text{ ft.}} = \mathbf{14 \text{ mph}}$$

Thus the formula for a two-cylinder double-acting engine that relates velocity in mph to wheel diameter in ft. and power strokes per min. is:

$$V = \frac{\pi D_W N\,(60)}{4\,(5{,}280)} = \frac{\pi D_W N}{352}$$

Traction power (P_T) converted to put V in mph is

$$P_T = \frac{TV\,(5280)}{33{,}000\,(60)} = \frac{TV}{375}$$

If no power is lost between the cylinders and the wheels, then the traction force at 14 mph must be:

$$T = \frac{375\, P_T}{V} = \frac{375\,(77)}{14} = \mathbf{2{,}050 \text{ lbs.}}$$

height of the chimney had to be restricted to 14 ft. above the rails (to pass under bridges, etc.) so that the natural draft was very small. A much greater draft is essential to bring the fire to the intense heat necessary to produce the quantity of steam and its high pressure. This was done by the blast pipe, located in the center of the chimney and into which the waste steam rushes, carrying air with it and causing the powerful draft through the tubes and in the fire. Thus the fire is not blown by forcing air into it but by sucking air from the flues. "The introduction of the steam blast for urging the fire and of the tubes for conveying the heated air through the water, are the principal causes of the great power of the present locomotives."[10] By such means, Stephenson designed his locomotive and thereby created a new industry and a new public facility.

Motion on Rails

Early in the nineteenth century people doubted that vehicles could carry heavy loads up inclines by traction: by smooth metal wheels running on smooth metal rails. George Stephenson had proven doubters wrong and by midcentury the numerical explanation seemed clear enough. In his treatise on the Stephenson locomotives, Marshall explained how the cylinder power got transferred through the driving wheels to the tracks.

The Stephenson locomotive of 1835 as described by W. P. Marshall had two double-acting cylinders each 12 in. in diameter and 18 in. long, with a pressure of 47.5 psi when operating at 14 mph.[11] The driving wheels were 5 ft. in diameter and they made 1 rev. for 2 power strokes of each cylinder. Thus 1 rev. of the driving wheels corresponds to 4 power strokes (2 per cylinder). The total power produced by the 2 cylinders was

$$\text{Hp} = \frac{47.5 \times 1.5 \times (113.1) \times 314}{33,000} = 76.7 \text{ Hp}$$

At this speed and with this Hp, the engine can pull along a level incline a total load R equal to the tractive force T exerted by the driving wheels on the rails. The tractive force times the velocity of the engine V gives a measure of the power where T in lbs. and V in mph can give Hp.

$$\text{Hp} = \frac{T \times V \times 5,280}{33,000 \times 60} = \frac{TV}{375}$$

and thus in the 1835 locomotive, the tractive force

$$T = \frac{375\text{Hp}}{V} = \frac{375 \times 76.7}{14} = 2,050 \text{ lbs.}$$

For the same power, if the velocity increases, the tractive force must decrease; or if the tractive force increases, as it must on an incline, the engine can keep the same power only if the velocity decreases.

Marshall reported that the locomotive could pull a total weight $W = 220$ tons on level tracks with the tractive force of 2,050 lbs. The train needs part of this force to overcome the friction of the wheels against an imperfectly aligned track. The engine needed very little force to overcome air resistance, or *drag*, compared to the thrust needed by Fulton's steamboat in water. The drag in water is much greater because water's density is more than 800 times larger than the density of air at sea level. Fulton based his power requirements exclusively on the thrust needed to overcome drag, whereas in the locomotive drag is far less important at comparable velocities. For example, at 14 mph, the drag of Stephenson's locomotive would have been about 18 lbs. This drag would require a thrust $T = 18$ lbs. and a power

$$Hp = \frac{18 \times 14}{375} = 0.67$$

As the power required, according to Marshall, was 77 Hp, we can see that the drag is negligible and nearly all the thrust is used to pull the train against track irregularities, inclines, and wheel friction.

The traction T is roughly proportional to 0.2 times the weight on the driving wheels, which implies that for this Stephenson engine, slightly more than 5 tons came on the driving wheels so that 0.2 (10,250 lbs.) = 2,050 lbs.[12]

Stephenson, Brunel, and the Gauge War

Robert Stephenson designed not only machines but also major structures, including the 1850 Britannia Bridge, the first tubular metal bridge. This structure had two main spans of 460 ft. each and carried the trains inside the tubes made of a floor, two vertical walls, and a closed top slab all of wrought iron.

In 1833 Robert Stephenson was appointed engineer-in-chief for the London and Birmingham Railway; he moved from Newcastle to London to supervise the survey although retaining direct oversight of his engine manufacturing company in Newcastle.[13] In the same year on March 7, 1833, I. K. Brunel (1806–1859) was made chief engineer for a new railway to run between London and Bristol. Brunel had worked for his father, Marc Brunel, on the latter's Thames tunnel in London but, having nearly

Figure 6.4 Isambard Kingdom Brunel (1806–1859)

died in an accident, was sent in 1829 to Clifton, near Bristol, to recuperate. By chance Brunel was there just at the time of the design competition for a bridge to cross the Clifton Gorge.[14] Although he had never before designed a bridge, Brunel entered and won the competition in 1831 with a suspension bridge spanning 830 ft., the longest spanning bridge in the world.[15] The Bristol riots of October 1831 stopped construction, and the bridge was not completed until 1864, five years after Brunel's death.[16]

Brunel remained in Bristol, became engineer to the dock company for the improvement of the Floating Harbour, and at age 27 got the job as engineer for the Bristol to London railway. He named it the Great Western Railway, expressive of his grandiose ideas for new works that through his boats would eventually link London to the United States and Australia by steam-powered vehicles.

Work began on the Great Western in 1835 after Brunel had convinced the Board of the Railway on September 15, 1835, to accept his own gauge of 7 ft.[17] Hence the broad gauge began in direct competition with Stephenson's narrow gauge. Brunel argued that the wider gauge would lead to greater speeds, smoother rides, and less maintenance. The gauge war had begun.

The Great Western construction began in London near Paddington and was completed to Maidenhead by 1838. On June 4, 1838, that part of the line was officially opened. Brunel designed a remarkable two-span brick arch at Maidenhead with 128-ft. spans and rises of only 24.5 ft. It was so flat that critics claimed it would never stand, but it still does after more than 150 years of service. It was opened on July 1, 1839, and the rail line opened for the public on March 30, 1840, to Reading. Finally, in June of 1841 the line was completed from London to Bristol (118 miles), which the fastest train could make in a little more than 4 hours at a rate of about 28 mph.[18]

Meanwhile Brunel had hired, in August 1837, a young locomotive designer, Daniel Gooch, who became his chief machine engineer and ordered locomotives for the broad gauge in late 1837. One of these was the *North Star* built by Robert Stephenson and Company. It began operating in January of 1839.[19]

On May 1, 1844, the broad-gauge line between Bristol and Exeter opened thus permitting travel directly from London to Exeter in about 5 hours, a distance of 194 miles or a speed of about 39 mph. The next year a broad-gauge line was completed to Gloucester, and there it came face to face with the narrow-gauge line of Robert Stephenson.[20] By 1845 it was clear that Britain could not have two different gauges for its main-line trains. As a result, on July 9, 1845, following a motion by Cobden, a royal commission was appointed to investigate the problem. During the investigation Brunel suggested a speed trial to show which gauge permitted faster transportation, and such trials were organized and run in late 1845. The broad-gauge trains proved faster but still the gauge commissioners in their report recommended that the narrow gauge be the standard, enforced by law. They gave two major reasons: First, it was much easier to convert broad-gauge lines to narrow-gauge lines than the reverse and, second, the total length of broad gauge in Britain was only 274 miles in 1845 compared to 1,901 miles of narrow gauge. The Gauge Act of 1846 put into law their recommendations and the broad gauge was thereby doomed to extinction.[21]

In spite of this defeat Brunel continued to design broad-gauge lines until his death in 1859. In 1847 he ordered six new locomotives including the *Sultan*, all with 8-ft. driving wheels. Nevertheless the broad gauge was gradually replaced by narrow gauge until it finally disappeared at the end of the nineteenth century. But many of Brunel's great bridges still stand, especially the 1859 Saltash Bridge, completed in the year of his death at a cost substantially less than Stephenson's Britannia Bridge.[22]

British versus American Railroad

The railroad came to the United States almost immediately after its initial success in Britain. There were no steam-driven lines in the United States in 1829, when the Rainhill trails publicized the new transportation in Britain, but by 1837 there were 1,450 miles of railroad in the United States as opposed to 1,600 in Britain in 1841. Railroading began seriously in the United States in 1829.

Just two months after the *Rocket*'s triumph at Rainhill, the Honesdale Railroad in Pennsylvania bought a locomotive from Robert Stephenson, the *Stourbridge Lion*, and that same year the Delaware and Hudson Canal Company imported four locomotives from England.[23] The railroad fever quickly accelerated from the eastern seaboard, partly stimulated by the immediate success of the Erie Canal, built between 1817 and 1825, a total of 365 miles. Philadelphia merchants realized that they needed a route west through Pennsylvania to compete with the New York State canal, so they formed the Philadelphia and Columbia Railroad in 1831. It used portage (hauling cars over the mountains by hemp rope), and it was on this line that John Roebling would later show the superiority of his iron-wire rope.

In 1834 steam locomotives appeared on the line and in 1832 Matthias Baldwin (1795–1866), an engraver and machinist, turned to building locomotives in Philadelphia. By 1838 there were in the United States about 350 locomotives, of which only about 78 were made abroad (predominantly in Britain). During the 1840s the U.S. locomotives began to differ sharply from the British designs. This "distinctive American style of locomotive—long, rakish, elegant if a bit spindly—emerged as a result of . . . the ideas of many builders, reflecting perhaps in its developed form an American point of view toward mechanical construction in general."[24]

By the 1850s the U.S. locomotives began to be sold abroad; British and U.S. engineers began to design and build railways overseas. In 1840 there were 3,300 miles of track in the United States (about as many miles as there were canals), but by 1850 the figure had risen to 8,900 miles. By 1860, stimulated by the federal government, which gave Midwest land grants to railroad companies, there were 30,000 miles of railways.[25] The railroads stimulated industry by requiring huge quantities of iron and then steel as well as coal; they basically transformed agriculture, allowing farmers to specialize and sell goods at great distances. Railroads became political forces and began to change people's perceptions regarding distance and time.

The difference between British and U.S. railroads stemmed from at least three factors. First, the British lines went between developed industrial cities like Liverpool-Manchester, London-Bristol, and London-Birmingham, whereas U.S. lines went from the developed East Coast out to the sparsely settled Midwest and then the Far West. Second, the U.S. railroads had to cross high mountains and wide rivers, which required large bridges, sharp curves, and steep inclines. Third, the lack of labor and the imperative for long rail lines in the United States led to lighter structures and machines less well suited for comfortable passenger travel than those short, straight, and smooth roadbeds in Britain.

Thus, in the United States there were developed a host of new bridge forms for rail traffic and a distinctive style of locomotive. By the time of the centennial, one could identify a maturity in these two styles: the massive and rigid British locomotives as exemplified by Robert Stephenson's designs and the light and flexible ones designed first by Baldwin and later by John C. Davis.[26]

To fit the more difficult U.S. terrain John Jervis devised as early as 1831 the swivel or bogie truck, consisting of four small wheels at the front of the locomotive that could swivel independently of the driving wheels and thus allow the long locomotive to make much sharper turns. In

FIRST RAILWAY TRAIN IN PENNSYLVANIA.

Figure 6.5 Portrait of Matthias Baldwin and the first railway train in
Pennsylvania

addition, Isaac Dripps invented the cowcatcher, a triangular metal front to
the locomotive to sweep away obstacles as the train traversed the long
stretches of unattended railway in the country. The locomotives also were
more flexible than their heavier, more rigid British counterparts. In addition, the driving wheels, usually four or six coupled together, were smaller
than the two or four British driving wheels. The smaller driving wheels
made for greater power but less speed, a necessity for the long, mountainous routes of the U.S. lines.

The *Lancaster* locomotive of 1834 by Matthias Baldwin was the
first one to operate successfully on the Philadelphia and Columbia Railroad. Baldwin had visited the Mohawk and Hudson Railroad in 1833 to
study J. B. Jervis's designs, which so impressed him that he built only
these designs. They maneuvered much better than the rigid British locomotives but they had less traction.[27] The *Lancaster* (see figures) had two
double-acting cylinders with $D = 9$ in., ($A = 63.6$ sq. in.), $L = 16$ in. (1.33

Baldwin's Lancaster
Locomotive of 1834

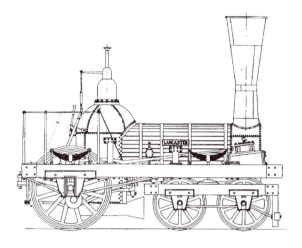

Baldwin's steam locomotive, the *Lancaster,* built in 1834

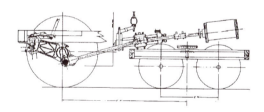

Running gear of the *Lancaster*

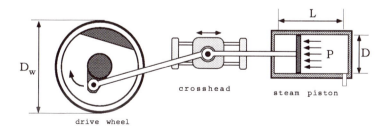

Diagram of the running gear

Engine Power

The engine power (P_E) from each of the two driving cylinders depends upon pressure of steam (P), length of stroke (L), area of cylinder (A), and number of strokes per min. (N).

$P = 50$ psi
$L = 1.33$ ft. (each power stroke)
$A = 63.6$ sq. in.
$N = 187.5$ power strokes per min.

$$P_E = \frac{PLAN}{33,000} = \frac{50\,(1.33)\,63.6\,(187.5)}{33,000} = 24 \text{ Hp/cylinder}$$

for two cylinders, $P_E = 2\,(24) = \textbf{48 Hp}$

Traction Power

Traction force (T) for iron wheels on iron rails is estimated to be $0.2W$ where W is the total weight on the driving wheels. For the *Lancaster,* this weight is 6,000 lbs. Thus, traction force (pulling force of the locomotive) is

$$T = 0.2\,(6,000) = \textbf{1,200 lbs.}$$

Velocity (V) of the driving wheels with diameter (D_W) of 4.5 ft. and strokes of 187.5 per min. for each cylinder or 375 strokes (N) per min. for the pair of cylinders is

$$V = \frac{\pi D_W N}{352} = \frac{\pi\,(4.5)\,375}{352} = \textbf{15 mph}$$

Traction power (P_T) is

$$P_T = \frac{TV}{375} = \frac{1,200\,(15)}{375} = \textbf{48 Hp}$$

Apparently conservatively, Baldwin stated that the engine could produce 30 Hp only, but with a steam pressure of 120 psi (see White, pp. 269–272). Other engines at this time had steam pressures ranging from 47.5 to 80 psi. We have used 50 psi here.

ft.), the driving wheel diameter $D_w = 54$ in. (4.5 ft.), and $P = 120$ psi. The total weight on the two driving wheels was 15,000 lbs. and the engine power, therefore, was 116 Hp.

The tractive force supplied at 15 mph would have been

$$T = \frac{375\text{Hp}}{V} = \frac{375 \times 116}{15} = 2,900 \text{ lbs.}$$

whereas the estimated tractive force based upon 0.2 times the weight on the driving wheels gives

$$T = 0.2 \times 15,000 = 3,000 \text{ lbs.}$$

indicating the reasonableness of the rule-of-thumb value of 0.2 in this case.

Early U.S. Rail Lines: New York City

The connection between New York City and New Jersey by steam-powered vehicles began with Fulton's steamboat and received stimulus in 1831 with the founding of the Mohawk and Hudson Railroad to compete with the Erie Canal, then in full operation. By 1839 a steam-powered railroad service was opened between South Amboy, New Jersey, and Camden; steamboats then connected South Amboy to Manhattan and Camden to Philadelphia. A Philadelphia-to-Trenton railroad was constructed by 1834 except for the Delaware River crossing, which was made in 1836 by a railway bridge. Thus, by 1839 one could go from Philadelphia to New York City in 4.5 hours by rail and ferry. The Pennsylvania Railroad completed in 1852 its main-line route between Philadelphia and Pittsburgh and began to develop, with other rail lines, a route to Chicago. By 1871, the Pennsylvania had consolidated its routes to the New Jersey side of the Hudson.[28]

Meanwhile in New York, Cornelius Vanderbilt was putting together a railroad empire between 1857 and 1869, by which time he controlled the New York Central Railroad and began to develop a route to Chicago through New York State, thence to Cleveland and west.

Thus began the intense competition between the Philadelphia-based Pennsylvania Railroad and the New York–based Vanderbilt lines. Pitted against the famous and flamboyant Vanderbilts was a Philadelphia engineer of a quite different character, whose name may have been temporarily lost in the late twentieth century. Moreover, this engineer possessed ideas about building, operating, and maintaining a technological system, also temporarily lost to the modern culture, much to its present grief.

J. Edgar Thomson and the Pennsylvania Railroad

By the end of the Civil War, the Pennsylvania Railroad had become the world's largest transportation company with more than 6,000 miles of track by 1873 and 8 percent of the nation's railroads.[29] More than

Figure 6.6 J. Edgar Thomson (1808–1874)

anyone else, the man responsible for building this empire was J. Edgar Thomson (1808–1874), who had joined the company as its first chief engineer in 1847 and was president from 1852 until his death in May of 1874.

In 1975 *Fortune* magazine selected the first 19 people for its newly established Business Hall of Fame. It did not include the most famous railroad financiers E. H. Harriman, James J. Hill, and Commodore Vanderbilt. But Thomson was named along with Henry Ford, Thomas Edison, and J. Pierpont Morgan. In describing Thomson, *Fortune* wrote about the impressive record of continual profits derived from his ability to cut costs while maintaining high-quality, reliable service and, best of all, "no scandal touched this man."[30]

> He built a great railway located in the industrial heartland of the United States, organized a superbly efficient administration to oversee the

daily operating details, groomed several talented men as successors, and left a corporate entity that proved fiscally stable and highly profitable. In fact, he was much too successful. The traditions he established, a strong emphasis on technical knowledge in the front office, promotion from within, and highly structured operating departments, all served the road well in its early years, but those truths were later chiseled in stone by heirs trained at his knee until the corporation became so conservative that it was no longer able to adjust to the changing times.

Like Thomson himself, the PRR's greatest strength was always in being a technological pacesetter. The sight of K-4 Pacifics charging up the 1.8 percent grade at the Horseshoe Curve trailing long lines of Tuscan red coaches; the Raymond Loewy streamlined GG-ls humming along the catenary at over 100 miles per hour, bringing the Broadway Limited in on time; early experiments with electric, gasoline, and diesel locomotives; the rapid postwar dieselization; the new centralized traffic control; and 250,000 freight cars bearing the line's familiar keystone illustrated that Thomson's spirit was alive and well on the road.[31]

Thomson, born 10 miles south of Philadelphia into a Quaker family, began work as a civil engineer in June 1827, surveying for rail lines in eastern Pennsylvania. In 1830 and 1831, he worked as chief engineer of one section for the Camden and Amboy Railroad in New Jersey, and from 1831 to 1834 he traveled and worked on other rail projects around Philadelphia. His reputation grew rapidly and on October 11, 1834, he received an offer to be chief engineer for the Georgia Railroad. Over the next 13 years he built and operated what was then the "longest railway in the world under one management."[32]

Meanwhile back in Pennsylvania the competition of New York's Erie Canal of 1825 was increasing New York City's dominance over Philadelphia at an alarming rate. In 1820 the two cities had almost the same population, but by 1850 Philadelphia had slipped to only about two-thirds the size of New York. Pennsylvania had succeeded in building its own railroad and canal over the mountains to Pittsburgh by 1834, but it was much more costly, slower, and could not compete with the Erie. Then when the Pennsylvania state legislature in 1845 approved a franchise for the Baltimore and Ohio (B&O) Railroad to go via Cumberland to Pittsburgh, Philadelphians felt deeply threatened and moved to form a consolidated rail line directly from their own city west.

The rail line was incorporated on April 13, 1846, and the state legislature approved both the new Pennsylvania and the B&O lines west but stipulated that the latter franchise would be void if the former could

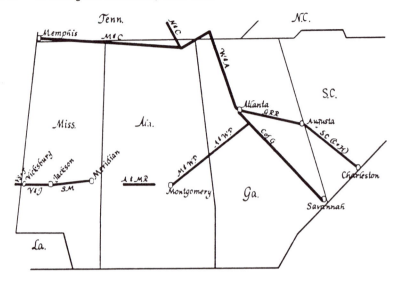

Figure 6.7 Map of the Georgia Railroad (c. 1847)

raise $3 million by July 30, 1847. Philadelphians were not able to organize formally until early 1847 and then quickly looked around for an engineer capable of designing and beginning construction with sufficient speed and cost control to convince investors to supply that large sum in only a few months.[33]

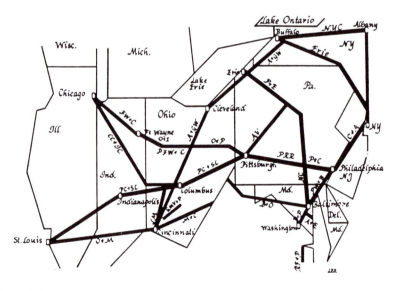

Figure 6.8 Map of the Pennsylvania Railroad (c. 1860)

By April 9, 1847, they had selected Thomson, who accepted the following week, came north immediately, and by June had the construction so well under control that the money was raised and the B&O charter declared null and void by the governor on August 2. So impressively did Thomson complete construction in the mountainous terrain that the line opened in 1850 with only the 36-mile Portage Railroad left as an impediment to straight rail service to Pittsburgh.[34]

While in Georgia, Thomson was asked in 1845 to rename the junction of the main rail line through Decatur to the state road at the tiny village of Marthasville. Thomson, in his usual terse manner, replied, "Atlantic, masculine; Atlanta, feminine, a coined word but well adapted." He thus named the major city Atlanta, and once again in central Pennsylvania when the line reached a junction in 1850 he named it Allatoona after a pass in Georgia; it was later shortened to Altoona.[35]

Just west of Altoona, Thomson designed the Horseshoe Curve, a sweeping switchback that carries heavy trains up over the mountains. The curve was part of the most difficult section that by 1853 had eliminated the need for the portage and finally made the Pennsylvania Railroad fully competitive with the Erie Canal.[36] The year before on February 3, 1852, the board had elected Thomson its president. He had risen to power through his engineering skill and an ability to manage both the construction and operation. Now he would be confronted with the need to operate in the world of high finance. Unlike the Vanderbilts and Harrimans, Thomson ran the railroad without himself having a primary financial interest in it. He did not view his railroad as a financial object to be manipulated but as a work of engineering to be carefully maintained and run at a profit.[37]

Thomson had reorganized the railroad by 1857 in a pattern that

> was eventually copied all across the country. Thomson's thought processes lent themselves to creating organizational solutions to complex problems; his engineering background trained him to think logically, to recognize cause and effect relationships, and to create mechanisms that operated with a minimum of friction.[38]

After the Civil War, Thomson poured millions into the maintenance of his line. He put profits back into the physical plant.

> On a more personal level Thomson lacked the consuming, driving avarice of a Gould, Vanderbilt, Carnegie, or Rockefeller; although he spent his entire career at least indirectly in the pursuit of profit, in keeping with his Quaker background he appropriated, in distinct contrast to his more flamboyant contemporaries, only an infinitesimal portion of his

income for personal use. He displayed as much enthusiasm for building and creating, as he did in collecting the rewards. He would not differentiate between his financial talents and his technical abilities. Each augmented and limited the other. This balance, increasingly unique in the postwar world, where specialization was becoming the norm, helped make him an important constructive force in mid-nineteenth century American economic development.[39]

Yet in no business or general history of the United States are Thomson and his work mentioned as a contrast to his "robber baron" contemporaries.

7 Henry, Morse, and the Telegraph

Railroad Collisions

Just before J. Edgar Thomson took over the Pennsylvania Railroad, a rival line to the north opened its route from the Hudson River to Lake Erie. On June 13, 1851, the New York and Erie Railroad celebrated its opening with a two-day train extravaganza led by President Millard Fillmore; his aged secretary of state, Daniel Webster; and numerous other political figures.[1] Onboard as well was Charles Minot, recently hired as the railroad's general superintendent. As the Erie developed its service through the summer of 1851, Minot rode the line personally, getting to know practically everyone who lived or worked near the newly finished road. One somewhat eccentric character he met was Ezra Cornell (1807–1874), a farmer turned telegraph promoter who was stringing lines near the Erie tracks at Goshen, about 60 miles from New York City.

Minot asked Cornell to install wires along the eastern section of the Erie line. On September 22, 1851, Minot got his chance to prove something big. He was travelling west on the single-line tracks of the road when they reached the small town of Turners, 47 miles from New York City. There were strict rules in those days that required trains to wait on side tracks until scheduled trains going in the opposite direction passed before they could proceed. The eastbound train was late coming to Turners so Minot's train had to wait. But Minot had no intention of waiting.

He telegraphed to Goshen 14 miles to the west, found out that the eastbound train had not yet appeared there, and ordered the stationmaster to hold up the eastbound train when it arrived. He then commanded the locomotive driver to forge ahead, which the driver strongly refused to do, jumping off the train. Minot took the throttle and went on to Goshen and eventually, after further telegraphing, all the way to Port Jervis, where the eastbound train finally appeared.[2]

Minot had saved more than an hour, but of far greater importance, he had proven the value of a major innovation: communication by electricity to control the traffic on single-line railroads. Schedules could be far more reliably followed and serious accidents avoided. The electric information age had begun.

The Early Telegraph

Communication at a distance had always been vital, especially in wartime. The light in Old North Church set off Paul Revere in 1775, fire beacons helped warn England of the approaching Spanish Armada in 1588, and smoke signals are a familiar part of western history. Modern systematic communications, however, did not start until the late eighteenth century.

In France the Chappe brothers established in 1794 a 145-mile communications line using telescopes and a mechanical semaphore code. The messages were sent visually from tower to tower with people at each tower acting as repeaters. The brothers coined the word *telegraph*, and the towers had to be high enough, often on "telegraph hills," to permit unobstructed sight.[3] By the 1840s there were numerous such lines in Europe, including one of 465 miles between Berlin and Koblenz (1833), and the French had by 1844 a network of 533 stations with 3,100 miles of these early optical lines.

Meanwhile, developments in electricity raised the possibility of wireline telegraphy thanks to Volta's invention of the voltaic battery by 1800, Oersted's 1820 discovery of the effect of an electric current on a compass needle, and Sturgeon's 1825 invention of the electromagnet. C. F. Gauss and W. Weber had built a practical electric telegraph by 1833, and by 1837 C. Wheatstone and W. Cooke had shown a working 1-mile telegraph in Great Britain.[4]

Volta, Ampère, and Ohm: The Electric Circuit

At the start of the nineteenth century, electricity was an obscure but intriguing mystery; by the century's end it would be a prodigious and powerful public force. In 1800 Alessandro Volta (1745–1827), a professor at Pavia, Italy, developed a battery consisting of a row of cups filled with acid or brine. Each cup had a copper plate and a zinc plate immersed in it, with each copper plate connected by metallic wire to the zinc plate in the next cup. At the two end cups there extended wires from one copper plate at one end and from one zinc plate at the opposite end. When Volta connected these two wires to make a circuit (a connected loop), then a current I flowed through them because of their potential difference, or voltage (named after Volta), V. The battery has a potential energy somewhat like the water behind a dam has a potential energy. Chemical reactions between the brine, or electrolyte, and each of the battery terminals, or electrodes, result in a condition where electrical charges have greater energy (more strictly, electrostatic potential energy) at one terminal than at the other. This is the voltage, or potential difference between two terminals, and is measured in volts.[5]

In 1827 a German physicist, George Simon Ohm (1787–1854), demonstrated that the size of the current flowing through the circuit depended upon the size of the voltage and also upon the resistance R within the circuit. The unit of resistance is now called the ohm and the unit of current, the amp, after the French scientist André Marie Ampère (1775–1836). The resulting equation, voltage = current × resistance, is called Ohm's law and is fundamental to electrical engineering:

$$V = IR$$

The resistance depends upon the cross-sectional area A, length L, and material of the wire in the following way:

$$R = \frac{\rho L}{A}$$

where ρ is the resistivity of the material. Copper, a good conductor, has a low resistivity, about 0.67×10^{-6} ohm-inches (ohm-in.) so that where $L = 1,000$ ft. (or 12,000 in.) and $D = 0.1$ in. ($A = \pi D^2/4 = 0.00785$ sq. in.)

$$R = \frac{0.67 \times 10^{-6} \times 12,000}{0.00785} = 1.02 \text{ ohms}$$

or about a 1-ohm resistance. If such a wire were connected to a 2-volt battery, the current in it would be

$$I = \frac{V}{R} = \frac{2}{1} = 2 \text{ amps}$$

In 1820, Hans Christian Oersted (1777–1851) observed the movement of a magnetic needle when located close to a wire conducting electricity. Ampère provided a theory for this electromagnetic effect and Joseph Henry (1799–1878), perhaps the United States' first great modern physicist, used this effect a decade later to construct electromagnets.[6]

Henry and the Working Telegraph

Henry began his work as a jeweler's apprentice; moved on to be an itinerant teacher of grammar around Albany, New York; and in the early 1820s served as an assistant to Dr. T. Romeyn Beck, a doctor of medicine and headmaster of the boys school, Albany Academy.[7]

As Beck's assistant, Henry prepared demonstration experiments in physics and chemistry and also continued his teaching as private tutor to the children of General Stephen Van Rensselaer (founder of Rensselaer Polytechnic Institute) as well as to the young Henry James, the father of William James the philosopher and Henry James the novelist. In addition, Joseph Henry proved to be an excellent surveyor, running the line for a new road between West Point and Lake Erie in the year following the

Figure 7.1   Joseph Henry (1799–1878)

opening of the Erie Canal. He was offered other civil engineering jobs but, in spite of a high salary, he declined them all to accept a position offered by Dr. Beck as professor of mathematics and natural philosophy at the Albany Academy. In the spring of 1826 Henry began his life as a teacher and scholar of science.[8]

That summer on a trip to New York City, Henry saw an electro-magnet designed in 1825 by William Sturgeon (1783–1850). It was a new invention and Henry resolved to build one himself. Back in Albany he worked to design a far more powerful one and, stimulated by an 1830 paper appearing in Scotland, Henry published his first major paper showing how a battery with a small voltage could produce a large magnetic force.

This demonstration would be the key to a practical telegraph in which a small battery could activate a distant magnet with enough electro-magnetic force to attract and then repel a small iron rod whenever the circuit were opened or closed. The problem, as Henry realized, was to get as much current as possible with as little voltage. Henry followed Sturgeon's design by wrapping the current-carrying electric wires around an iron-bar core bent in the form of a horseshoe. But he recognized that the strength of the magnetic field would be raised not only by increasing the current but also by increasing the number of turns the wire made around the core.[9] In experiment 15 from his 1831 paper, Henry described how his 21-lb., horseshoe-shaped, iron-core magnet lifted 750 lbs. powered only by a single zinc-copper cell battery. He emphasized the efficiency of his design, in which the energy produced by the single-cell battery allowed him to create, in his words, "probably the most powerful magnet ever constructed."[10]

Henry made his electromagnet by bending a 20-in. bar of soft iron into the form of a narrow horseshoe 9.5 in. in height. The bar had a cross-sectional area A_c of $2 \times 2 = 4$ sq. in. The ends were ground flat to allow an armature of the same material to fit flush against the bottom. With the iron armature in place, the magnetic field lines follow the closed path (within the iron) having a length L_c of about 23 in. The horseshoe was wrapped with 9 coils, each wound from 60 ft. of copper wire 0.045 in. in diameter (see the following figure). The resistance of the wire in each coil was about 0.3 ohms.

For Henry's zinc-copper battery, providing about 1.0 volt, the current flowing in each coil was 3.3 amps. Each coil consisted of about 80 turns around the 2-in. square iron core. With all 9 coils connected to the battery (as in Henry's experiment 15), the horseshoe was excited by a total magnetic potential:

$$IN = 3.3 \times 9 \times 80 = 2,376 \text{ ampere-turns}$$

Joseph Henry's Magnet

SB7A HERE

Frame for testing strength of electromagnet

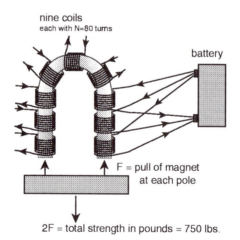

nine coils
each with N=80 turns

battery

F = pull of magnet
at each pole

2F = total strength in pounds = 750 lbs.

Circuit model for nine coils in parallel

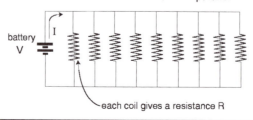

battery
V

I

each coil gives a resistance R

Henry knew that by having the total 720 turns split into 9 separate coils, he could get a far higher current through the wires with the same single-cell battery of only 1 volt. This is so because he put the 9 circuits in parallel rather than in series (see the figures below).

The magnetic potential introduces a magnetic flux into the core in much the same way that a voltage produces the current in a wire. In turn that flux gives the magnet its strength to lift a weight. That strength or force for Henry's magnet was 750 lbs.[11]

In the parallel circuit, 1 volt produces 3.3 amps in each circuit; but in the series circuit the 1 volt battery produces only 0.37 amps in the single circuit and hence would give a total magnetic potential of only 266 ampere-turns. This would have given a force of only about 100 lbs. The crucial insight Henry developed in his study was that the force was an increasing function of the number of ampere-turns and that small voltages could create much larger currents only if the turns were divided into numerous branches connected in parallel to the battery.

Current in One Coil

Ohm's law, $V = IR$ **Resistance $R = \rho \dfrac{L}{A}$**

V = *The voltage* produced by the battery which we assume, for
   Henry's experiment 15, to have been 1 volt.

R = *The resistance* throughout each of the 9 circuits of length L.

L = 60 ft., *the length* for one wire going from the battery to the magnet,
   coiled around the iron core, and returning to the battery (one circuit).

A = 0.00159 sq. in. the *cross-sectional* area of the .045-in. diameter
   copper wire.

ρ = The *resistivity* of the wire material, which for copper is
   0.67×10^{-6} ohm-in.

Hence, the resistance in each of the 9 parallel circuits is

$$R = 0.67 \times 10^{-6} \frac{60 \times 12}{1590 \times 10^{-6}} = 0.303 \; ohms$$

and, from Ohm's law, we calculate the current as

$$I = \frac{V}{R} = \frac{1}{0.303} = 3.3 \; amps \text{ in each of 9 parallel circuits.}$$

In the fall of 1830, stimulated by reading a failed attempt by an English mathematician, Henry strung up more than 1 mile of wire around his classroom and, from a strong battery at one end, sent a current through the wire, activating a magnet at the other end. The magnet caused a pivoted steel bar to strike a bell. Henry had demonstrated "instantaneous" transmission of motion by electricity. He announced the idea in the July 1831 issue of *Silliman's American Journal of Science*.[12]

Henry's research in Albany gained him respect in the United States, and when the College of New Jersey decided to appoint a new professor of natural philosophy (physics), they consulted Benjamin Silliman

Magnet Strength

The strength of the electromagnet, F, depends upon the flux density *B* and the pole area *A*.

$$F = \frac{B^2 A}{72{,}130{,}000} \text{ lbs.}$$

The flux density B depends upon the number of turns of wire $(N = 9 \times 80 = 720 \text{ turns})$ coiled around a core of length $L_C = 23$ in., the current flowing through the wires $(I = 3.3 \text{ amps in each turn})$ and the material properties of the core:

$$\frac{IN}{L_C} = \frac{3.3(720)}{23} = 103 \text{ ampere-turns per in.}$$

The flux density varies with the quantity IN/L_C as well as with the materials; it must be taken from experimentally derived magnetization curves. For the soft iron used by Henry, we can estimate, taking standard curves for iron and steel with $IN/L_C = 103$,

$$B = 82{,}000 \text{ lines of flux per sq. in.}*$$

The pole area A is the cross-sectioned area of one end of the horseshoe core or $2 \times 2 = 4$ sq. in. Therefore, the total strength of both poles

$$2F = \frac{2(82{,}000)^2(4)}{7.213 \times 10^7} = 746 \text{ lbs.}$$

is about what Henry measured.

*For such curves see, for example A. L. Cook and C. C. Carr, *Elements of Electrical Engineering*, 5th ed., New York, 1947, p. 30. We have interpolated and picked the value that gives Henry's result.

Parallel Circuit for 9 Coils

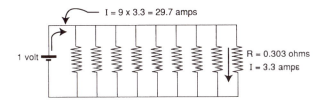

I = 9 x 3.3 = 29.7 amps

1 volt

R = 0.303 ohms
I = 3.3 amps

$$I = \frac{V}{R_1} + \frac{V}{R_2} + \frac{V}{R_3} + \frac{V}{R_4} + \frac{V}{R_5} + \frac{V}{R_6} + \frac{V}{R_7} + \frac{V}{R_8} + \frac{V}{R_9} = 9 \times \frac{1}{0.303} = 29.7 \; amps$$

$$\frac{IN}{L_C} = \frac{(3.3)(720)}{23} = 103 \; \frac{ampere-turns}{in.} \qquad B \approx 82,000 \; \frac{lines \; of \; flux}{sq. \; in.}$$

$$2F = \frac{2(82,000)^2 \, 4}{7.213 \times 10^7} = 746 \; lbs.$$

Series Circuit for 9 Coils

$I = \frac{1}{9 \times 0.3} = 0.37$ amps

1 volt

R = 0.303 ohms
I = 0.37 amps

$$I = \frac{V}{R_1 + R_2 + R_3 + R_4 + R_5 + R_6 + R_7 + R_8 + R_9} = \frac{1}{9 \times 0.303} = 0.37 \; amps$$

$$\frac{IN}{L_C} = \frac{0.37(9 \times 80)}{23} = \frac{266}{23} = 11.6 \; \frac{ampere-turns}{in.}; \quad B \approx 30,000 \; \frac{lines \, of \; flux^*}{sq. \; in.}$$

$$2F = \frac{2(30,000)^2 \, 4}{7.213 \times 10^7} = 100 \; lbs.$$

Henry got over seven times the strength by using a parallel instead of a series circuit for his electromagnet.

*Note that B is not linearly proportional to $\frac{IN}{L_c}$.

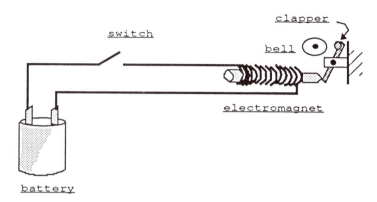

Figure 7.2 A schematic diagram of Henry's telegraph

of Yale, among others. Silliman stated that "Henry has no superior among the scientific men of the country—at least among the younger men," and Professor Renwick of Columbia announced that "He has no equal." Henry himself, in responding to inquiries from the college, asked "Are you aware of the fact that I am not a graduate of any college and that I am principally self-educated?"[13]

Such recommendations led to Henry's appointment in 1832. At Princeton he taught physics, chemistry, geology, mineralogy, and astronomy; moreover, he also taught Princeton's first course in architecture. He was so prized by the college that it gave him a full year's paid sabbatical in 1837 to travel to Europe where he met Faraday, Wheatstone, and Babbage, among others, and where he bought scientific instruments for use at Princeton.

While at Princeton he built a magnet that lifted 3500 pounds and he strung wires between his house (which he designed himself) and his laboratory so that he could telegraph his wife.

While at the college he came to be seen as the foremost scientist in America so that when James Smithson (1765–1829) left his fortune to found an institution in the United States devoted to "the increase and diffusion of knowledge," the nation turned to Joseph Henry to draft a plan for the new endeavor. The organizing committee then asked Henry to lead the new institution, where he began a new life on December 14, 1846.[14]

While at Princeton he had met Morse and in 1839 had encouraged him in his promotion of the telegraph. In 1857, after the telegraph had proven its immense worth and large litigations had begun, Henry recounted its history to illustrate clearly his pioneering role in its invention.

Figure 7.3   Joseph Henry's electromagnet and hammer

However, he was at pains to stress his complete lack of motivation to turn invention into innovation. He ended his deposition as follows:

> The results here given [of the invention of telegraph] were among my earliest experiments: in a scientific point of view I considered them of much less importance than what I subsequently accomplished, and had I not been called upon to give my testimony in regard to them, I would have suffered them to remain (without calling public attention to them) a part of the history of science to be judged by scientific men who are best qualified to pronounce upon their merits.[15]

Such a modest view was not typical of the man most associated with the telegraph, Samuel Finley Breese Morse (1791–1872).

Figure 7.4 Joseph Henry's magnet

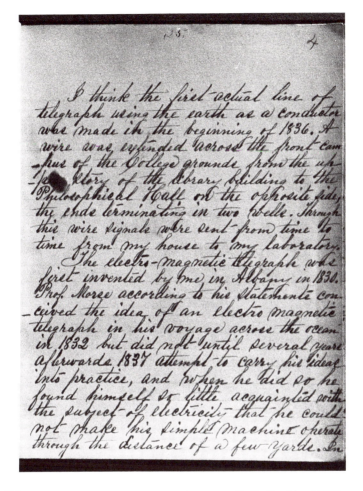

Figure 7.5 Letter written by Joseph Henry to S. B. Dod in 1878 describing his research while at Princeton. In it he tells of his invention of the telegraph in 1830. He had a telegraph line between his home and his laboratory in Philosophical Hall, which he used to tell his wife when he was coming home to lunch.

Morse and the Telegraph

All these advances made it possible for Samuel Morse to develop a practical telegraph in the United States. Just as the Newcomen engine was an essential predecessor to Watt's idea, so was the invention of a practical telegraph the necessary precondition for the overwhelming productivity of Edison.

Figure 7.6 Samuel Finley Breese Morse (1791–1872)

Morse graduated from Yale College in 1810 with a not distinguished record but a strong urge, as had Robert Fulton, to be a painter. Like Fulton before him, Morse went to London to study with Benjamin West. Unlike Fulton, however, Morse became perhaps the most prominent U.S. artist of his day. He became president of the National Academy of Design and professor of the literature of the arts of design at New York University. He was, thereby, the first professor of art in the United States.[16] Between 1810 and 1837 he tried without success to gain a financially secure life by painting, but he could not sustain such a career in the United States during that period. The country was as yet unready to support the fine arts in the way he expected.

Meanwhile Morse, who had been attracted to electricity ever since his college days, had begun to develop an idea for sending messages over long distances by means of wires. The idea for a practical design apparently occurred to him on shipboard while returning to the United States in 1832. At New York University in 1835 he set up a laboratory referred to as a studio in which he worked on telegraphy. Soon he had a few financial backers for the idea. Morse was encouraged by Joseph Henry, but the two eventually had a falling out largely because of "their different understanding of the relationships between science and technology. Henry believed that the role of science was to make discoveries and that a mature science opened the door to practical applications that would be achieved as soon as society was ready to put through the technological change. Inventors were 'men of action,' accessory to scientists, who were 'men of mind.' Morse much resented the view that invention and work with technology were things other and less than intellectual activity. He was convinced that all of his work was based upon thought."[17]

Morse and his backers tried to obtain a grant from the U.S. government as early as 1838, and finally did succeed in 1843 in getting $30,000 to build a telegraph line between Washington, D.C., and Baltimore. It was built within that budget and successfully operated on May 24, 1844, sending the words "What hath God wrought." A company was formed a year later, and the Western Union Telegraph Company consolidated numerous small companies in 1856. By 1865 more than 200,000 miles of telegraph lines were in service.[18]

The secret of Morse's success lay largely in his ability to visualize devices as part of a whole system (transmitter, lines, and receiver) and his invention of a simple code (the Morse code with minor changes was still used regularly during World War II) in which each letter was represented by a small series of dots and dashes. As had happened with Fulton, Morse's artistic ability to visualize and his motivation for business success combined to drive him toward a commercially useful design.

Invention and Innovation

How did Morse the painter become Morse the telegrapher? This transformation took place in a series of stages beginning with his move into the new Gothic structure completed in 1835 for New York University. Morse's studio was ostensibly for the teaching of art but in reality it served to house his electrical experiments. Morse's first stage sprang from his shipboard design of a full telegraphic system, which, however, had little detail. Indeed, when he tried to build a working model, his clumsy mechanical skill and his lack of knowledge in electricity and magnetism kept him from any workable system.

For the second stage, Morse received essential help from a colleague, Leonard G. Gale, a science professor who had once worked with Joseph Henry and who understood the results of Henry's research. Gale could see right away two conflicting facts: First, Morse had good ideas and, second, he did not realize that telegraphy required a much stronger battery to overcome long-line resistance. Moreover, Morse's magnets had only a few turns of wire and hence, along with his weak current, could produce very little magnetic force (flux ϕ). So Gale provided insights from science that the operation of the telegraph demanded much higher N and I. In early 1836, by increasing the wire turns from a few to more than 100 and by greatly increasing the battery voltage, Gale and Morse could send a signal over 10 miles of wire. Morse invited Gale to be his partner. The artist designer combined with the scientist designer, but they still lacked the third ingredient essential to turn invention into innovation: money.[19]

As if by design, financial support and more came from a student, Alfred Vail, just graduated from the university and enthusiastic over the telegraph. Vail's father, a wealthy ironmaker in New Jersey, agreed to finance full-scale testing and promotion. Alfred himself also brought substantial mechanical talent to the professors' design and scientific abilities. The improved system had its first formal demonstration at the Vails' ironworks in early 1838, encouraging the partners, now including Vail, to begin the next stage in the process of innovation: politics.

In February, Morse and Vail went to Washington, D.C., where Morse demonstrated the new system to President Martin Van Buren and his cabinet as well as to the House Committee on Commerce. Morse now began to get broader support in his new quest for a government grant to build a working system.[20] He also took a year to go to Europe in the hope of getting further support; it was a wasted time except for his discovery of Louis Daguerre, whose technique he brought back to New York, setting up a studio to teach the new art of photography. "Art is to be wonderfully enriched by the discovery," he wrote. "Our studios will now be enriched with sketches from nature which we can store up during the summer . . . and thus [in winter] have rich materials for compositions and an exhaustless store for the imagination to feed upon."[21]

Morse's major interest remained the telegraph, which in 1840 he patented, and in 1842 he again petitioned Congress for support. Finally on March 3, 1843, Congress granted Morse $30,000 to build a trial line between Baltimore and Washington. From design, to science, to finance, to politics, Morse finally got the chance to prove the merit of his system.

Ezra Cornell strung his wire and on May 24, 1844, the first message went through. Intelligence at a distance was a reality and would soon cover the continent.

Morse still hoped to sell his telegraph to the U.S. Congress and to leave the harrowing business world, but Congress refused. Morse then hired a lawyer, Amos Kendall, who had been postmaster general and a close associate of President Andrew Jackson. Kendall also failed to get the government to buy the system, so he founded a network of companies to construct and run lines privately. First came the Magnetic Telegraph Company, formed in 1845, to connect New York to Philadelphia; other lines quickly followed. Morse, as a major shareholder in this and other companies, became a wealthy man, retiring in the 1850s to a large house near Poughkeepsie and travelling to Europe to receive numerous honors and prizes.[22]

In spite of the immense success of his physical system, Morse's fame endured well into the twentieth century thanks largely to his code, which might properly be called the first modern information system. His objective had been to devise a simple means of communicating messages with as few errors as possible. Just as important was the idea that even when errors occurred the message would still be intelligible. Morse's dot-dash system satisfied these criteria and nothing better was put into wide use for a full century.

In the National Portrait Gallery there hangs a large painting entitled *Men of Progress*. The painter, Christian Schussele, showed 19 figures, including Joseph Henry, but at the center and shown most prominently is Samuel F. B. Morse, clearly in this view placed at the summit of achievement. Is that position merited by Morse or by any single figure? The answer to those questions characterizes much of the story of engineering during the nineteenth century.

We can observe at this point that Morse did deserve credit for the complete telegraphic system—including both hardware and software—even though he needed substantial help from Gale, Vail, Henry, and some political figures in Washington. He also needed to appear at the right time and place. Benjamin Franklin certainly possessed the mind to conceive of the telegraph but he lived before the Volta battery and the Henry electromagnet. Of course, Morse also needed someone like Ezra Cornell, who knew how to construct things and who recognized the need to string wires on poles rather than Morse's idea of putting them underground. But with all their help, none of these people would have done the telegraph without the central perseverance and drive of Morse. Morse, the central innovator,

Figure 7.7 Ezra Cornell (1807–1874)

put the system together and made it a commercial reality. Yale named one of its colleges for Morse, but Cornell went well beyond that.

In 1855 Ezra Cornell, with others, organized the Western Union Telegraph Company, in which he was for a while the largest stock-holder.[23] He then gave large sums of his wealth to found a university on the site of his farm in Ithaca, New York, following the 1862 passage of the Morrill Act.

This act provided the land grants to create national support for agriculture and engineering in each state. Cornell University opened in

1868 and quickly became a great university, with one of the most distinguished engineering schools in the country. The telegraph fortune was converted into education as the lines of wire went together with the lines of rail to connect the emerging states of the post–Civil War Union.

By 1884, Western Union had more than 142,000 miles of lines, 12,386 telegraph offices, and gross earnings of more than $19 million, of which $7.7 million were net profits.[24] Moreover, the telegraph combined with the railroad produced a major transformation in the nation's interconnectedness through the creation of "Railway Standard Time" on November 18, 1883. The telegraph made possible the synchronization of time that rationalized train scheduling and produced the time zones we still have in the late twentieth century.[25]

8 St. Louis versus Chicago and the Continental Railroads

War and Peace on Rails

In 1888, *Scribner's Magazine* began publishing a series of articles on the state of railroads in the United States. Published in book form the following year, the collection presented not only a descriptive review of the world's largest business, but also intimations of its immense social role throughout the country. In the introduction, Michigan Judge Thomas M. Cooley, the first chairman of the newly formed Interstate Commerce Commission, spoke of the railroad situation during the post–Civil War period as one "without system . . . so operated that the antagonism of managers, instead of finding expression in legitimate competition, would be given to the sort of strife that can only be properly characterized by calling it, as it commonly is called, a war." Cooley was at pains to emphasize how misleading the term *system* was when applied to the railroads of 1889, and yet felt that much of it was "marvelous" and "the wonder of the world."[1]

Cooley saw the *Scribner's* collection of 13 separately authored chapters as having permanent value, and viewed a century later as characteristic of the time, they indeed do have such value. There may have been a war into which Judge Cooley was asked to help bring peace, but the country as a whole was at peace and riding the steel lines to high prosperity and self-confident industrial dominance. As the first paper, continuing the military metaphor, announced, "the world of today differs from that of Napoleon Bonaparte more than his world differed from that of Julius Caesar; and this change has chiefly been made by railways."[2] Such a theme would later be stated poignantly, not confidently, by Henry Adams, whose brother Charles Francis Adams, Jr., then president of the Union Pacific Railroad, had written the eleventh chapter, which dealt with how to avoid railroad strikes.

In a section entitled "Railroads and Democracy," Thomas Curtis Clarke proclaimed that "the grand function of the railway is to change the whole basis of civilization from military to industrial." He observed that the money used in "maintaining the whole of Europe as an armed camp" is, in the United States, spent on building and maintaining railways. He further reminded his readers that without the railroads "the rebellion of the southern states could never have been put down, and two great standing armies would have been necessary." He went on to argue that the railroads, by permitting rapid transportation of food, free farmers from landlords and allow political power to move from the large landowner to "the owners of small farms, and the manufacturers and merchants." Railroads were thereby crowned as the means for democracy, for greater human happiness, and for recognizing "the engineer [who] should jointly take rank with statesmen and soldiers, and that no greater benefactors to the human race can be named than the Stephensons and their American disciples."[3]

One measure of the immense change wrought by Stephenson's disciples lay in the rise in power of the locomotives that made transcontinental travel possible in fewer than four days. In chapter 3 of the Scribner's book, M. N. Forney, editor of the *Railroad and Engineering Journal*, presents the data that explains the transformation. The quaint little engines from England had begun the changes but it was the immense machines made by companies like Baldwin in Philadelphia that finally broke the barriers of the Alleghenies, the Rockies, and the High Sierras.

By the 1880s the engine power reached 3,000 Hp and the ordinary velocity was up to 60 mph, with driving wheels of 5.5 ft. in diameter and with an available traction or thrust power of around 2,000 Hp, compared to the locomotive power in the early 1830s of 100 Hp.[4] This vastly increased power came to the wheels in the form of a traction or pull equal to about $0.2W$, where W is the total weight on the driving wheels, which for the 92,000-lb. "American"-type passenger locomotives of the 1880s reached about 65,000 lbs. Hence the total traction or pulling force would have been $T \approx 0.2 \times 65,000 = 13,000$ lbs. and the resulting power at 59 mph.[5]

$$\text{Hp} = \frac{TV}{375} = \frac{13,000 \times 59}{375} = 2,050 \text{ Hp}$$

"American" Passenger Locomotive, 1880

Engine Power

$P = 150$ psi

$L = 2$ ft./stroke

$A = 18^2 \, \pi/4 = 254.5$ sq. in.

$N = 1,200$ strokes per min. (600 strokes/min. per cylinder)

$$P_E = \frac{150 \times 2 \times 254.5 \times 1,200}{33,000} = 2,780 \text{ Hp}$$

Traction Power at the Rails

$W = 92,000$ lbs. for the locomotive weight

$W_D = 65,000$ lbs. weight on the four driving wheels, or 16,250 lbs. per wheel

$T \approx 0.2 \, W_D = 13,000$ lbs. traction (thrust) on the four driving wheels, or 3,250 lbs. per wheel

$D_W = 5.5$ ft. diameter of each driving wheel

$$V = \frac{\pi D_W N}{352} = \pi(5.5)\frac{1,200}{352} = 59 \text{ mph}$$

$$P_R = \frac{TV}{375} = \frac{13,000(59)}{375} = 2,050 \, Hp$$

Effiency, $\dfrac{P_R}{P_E} = \dfrac{2,050}{2,780} = 74\%$

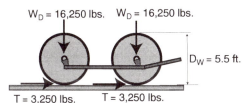

$W_D = 16,250$ lbs. $\quad$ $W_D = 16,250$ lbs.

$D_W = 5.5$ ft.

$T = 3,250$ lbs. $\quad$ $T = 3,250$ lbs.

Traction per driving wheel, which moves the train forward (to the right in this case)

Again, this meant about 2,000 Hp, which pulled large trains over the mountain ranges to make transcontinental travel by 1870 radically different than it had been before the Civil War, when covered wagons took a season to make it to California.[6]

The weight of 65,000 lbs., necessary to develop the traction of 13,000 lbs., was far too much for only two wheels—the steel would deform too much under such large loads. The maximum load in 1880s ranged between 12,000 and 16,000 lbs. per wheel and for that reason designers used more wheels. The typical large "American"-type locomotive of the time had four driving wheels, thus in our case giving about 16,000 lbs. per wheel.[7] Another result of these far more powerful locomotives was the lowering of freight rates: from 2.9 cents per ton-mile in 1865 to 0.718 cents per ton-mile in 1887. By 1889 about 10 percent of the wealth of all nations was in the railroad and they held between one-third and one-quarter of all capital investment in the world.[8] Such was the transforming power of the steam-driven locomotive. There was no more powerful image of nineteenth-century U.S. society, which was connected by 150,000 miles of railroads, more than the sum of all mileage in the rest of the world.[9]

Crossing the Continent

This immense network moved across the western states rapidly only after the Civil War ended. Legislation passed by Congress in 1862 and signed on July 1 by Abraham Lincoln stated that the line to the Pacific Ocean would be built by two companies: the Central Pacific from Sacramento, California, east and the Union Pacific from the Missouri River west. Work did not begin until after the Civil War.

The Union Pacific Railroad picked General Grenville M. Dodge (1831–1916) early in 1866 to be chief engineer for the line running from Omaha west to Utah. Meanwhile in California, four merchants came together to get control of the Central Pacific. Leland Stanford (1824–1893), a grocer, was president of the line and later Republican governor of California. Collis P. Huntington (1821–1900) became the chief agent in the East to raise funds and deal with politics while his former hardware store partner, Mark Hopkins (1813–1878), was treasurer. Charles Crocker (1822–1888), a gold miner who became a dry-goods merchant, served as the head of construction.

The two companies built at breakneck speed to secure as long a line as possible when they met. Extra lines meant extra government subsidy. Huntingdon tried to get President Andrew Johnson (1808–1875) to agree to a more easterly meeting while General Dodge, who in addition to being chief engineer was a congressman from Iowa, worked on his old army friend Ulysses S. Grant (1822–1885) to get the Union Pacific more line to the west. On April 10, 1869, the Congress intervened and established a lonely place in Utah called Promontory Point as the meeting place. On Monday morning of May 10, 1869, Leland Stanford arrived on a special train from Sacramento and Thomas C. Durant (1820–1885), vice-president of the Union Pacific, arrived in a bright Pullman car with General Dodge.[10]

California Governor Leland Stanford swung the hammer to pound in the last (Golden) spike on the first transcontinental rail line (he missed but Dodge succeeded). It was clear that the railroads were playing a central role in the settlement of the vast nation and that they were "paving the way for America's industrial leadership of the world."[11]

Chicago versus St. Louis

As the nation began to compete for industrial leadership in the world, its geographic parts competed among themselves for regional dominance, and none did so more intensely than the two major cities of the Midwest: Chicago and St. Louis.[12]

The development of Chicago up to the fire of October 8, 1871, was that of a boom town turning into a metropolis by means of transportation (the railroad) and the agricultural and livestock industry (such as farm equipment and slaughterhouses). It was a city with almost no history and a collection place for people on the move.

During this same early period, the principal competition to Chicago's dominance of the Midwest was St. Louis, a city that strongly contrasted with Chicago. St. Louis was much older, founded in 1764 by French-speaking settlers from New Orleans. Chicago was still little more than a village by 1840, comparative populations show:

	St. Louis	Chicago
1840	16,500	4,500
1850	75,000	30,000
1860	161,000	109,000
1890	313,000	298,000

St. Louis retained its tie to the South through its river commerce with New Orleans. In fact, it was the St. Louisan James Eads (1820–1887) who in the late 1870s constructed the jetties that permanently opened up New Orleans's badly silted channels to the sea. Chicago, meanwhile, built up elaborate ties to the East, especially New York City, through its railroad commerce. The Civil War badly disrupted St. Louis commerce by cutting off the North-South trade, whereas it encouraged Chicago's growth by accelerating East-West trade.

The steamboat symbolized St. Louis and its more easy-going, more elegant, and more self-contained civic attitude. It never developed such close ties to the East, with its immense capital potential for industrial development. Chicago, tying itself to the railroad, reached out to receive the great agricultural produce from the Midwest, processed it, and sent it on to the huge markets in the populous East.

A visitor from Britain wrote in 1870 that the growth of Chicago "is one of the most amazing things in the history of modern civilization." This perceptive woman reflected on the change in one lifetime from "the haunt of Indians" to the metropolis whose street railways carry annually 7 million passengers: "The log huts have made way for magnificent warehouses and palaces of marble; the little traders have become great merchants, some of them worth millions of dollars, and doing business on a scale of extraordinary magnitude."[13]

The cause of this growth was the railroad, coming from the east and spreading out to the west. This engineering network focused on Chicago, where commerce in grain and meat drew people of talent and investment for profit. Railroads shaped the city as they shaped the country. "Tracks, terminals, and other facilities altered the appearance of the city, while suburban stations became the centers of population beyond the municipal borders."[14]

Whereas only one railroad entered the city in 1850, by 1856 ten rail lines—parts of networks totaling 3,000 miles of track—had already made Chicago the world's largest railroad center. Then came the Civil War and the dominance of Chicago seemed confirmed. St. Louis offered, however, one last challenge that took the form of a bridge.

St. Louis versus Chicago: The Eads Bridge

Rapid westward expansion of the railroads after the Civil War had emphasized the need for a bridge at St. Louis. Rail cargo was being unloaded along the East St. Louis levee, loaded on ferries, and transported to the west shore where a similar process was carried out, all at the price of much delay.

Figure 8.1   James B. Eads (1820–1889)

U.S. Senator Gratz Brown from Missouri introduced and influenced the passage of a bill in Congress approving the St. Louis and Illinois Bridge Company's plans. His bill was so amended that the bridge could not be a suspension or draw and required a maximum opening of either two spans of 350 ft. or one of 500 ft., with a clear height of 50 ft. above the city directrix. Suspension bridge failures had ruled out their acceptance and railroads had lobbied successfully against a drawspan, which could delay traffic.

Senator Brown introduced his bill with the following prophetic statement:

> I want the structure when built to be one worthy of the great States it is to connect, of such ample capacity as will permit the freight of all the railway lines that may hereafter center at the great distributing point of the continent. Let it be built, too, for the ages, of a material that shall defy time, and of a style that will be equally a triumph of art and a contribution to industrial development.[15]

James B. Eads had begun working on his own plans for a bridge in 1866, and at a meeting of the St. Louis and Illinois Bridge Company on March 23, 1867, Eads was elected engineer-in-chief. At that meeting he presented his initial drawings for an arch bridge and the company decided to begin work under the existing charters as soon as possible despite opposition from Chicago interests. One of Eads's first moves was to hire as his assistant an experienced engineer, Colonel Henry Flad, then a member of the St. Louis Board of Water Commissioners.[16]

Eads had almost immediately to face the competition of Chicago because that city saw the bridge as a tangible symbol of the threat to its emerging dominance of the central part of the United States. This competition assumed the form of a rival bridge builder with the appropriate name of L. B. Boomer from Chicago, who formed his own bridge company, had his own design made, and proceeded to begin construction of a totally different bridge on the Illinois side of the Mississippi at East St. Louis. The idea of two radically different bridges meeting in midstream was too ludicrous to contemplate but as a negotiating position it had some force. Eads prevailed and got his arch bridge completed by 1874, but it was too late to stave off the power of Chicago.[17]

Railroads and Steamboats

When congressional authorization of the bridge in 1866 required one span of 500 ft. or two of 350 ft. and a 50-ft. clear height above the city directrix, the steamboat-chimney issue had been discussed, it being generally agreed that stacks would be lowered. When the arches were closed, however, steamboat companies operating out of St. Louis recognized the finality of the threat, not only to their smokestacks, but also to their profits. They recognized the threat of the railroad to their business.[18]

Smokestacks were later lowered well below the 50-ft. clear height of Eads Bridge with no adverse effects to steamboat operation, but the introduction of 20 railroads operating into St. Louis in 1873 was a different story. In mid-1873, John S. McCune, president of the Keokuk Northern Line Packet Company, and others protested to General William W. Belknap, secretary of war, that the bridge was a hazard to navigation. One may well wonder why the steamboat companies had waited so long to protest, for the bridge had been under construction almost six years. The reason lies in the fact that Belknap had only recently, by the Act of Congress of June 4, 1872, been granted authority for "the security and convenience of navigation, to convenience of access, and wants of all railways and highways crossing said river."[19] Belknap, a friend of the Keokuk Company, had been secretary of war since 1869 but had previously lacked authority over the Mississippi River navigation. As far as can be determined, Belknap had no financial connection with Keokuk.

The result of this protest to Belknap was a report from the Army Corps of Engineers condemning the bridge and leading to the recommendation approved by the chief of the Corps, General A. Humphreys, that it be torn down. His report was forwarded to Belknap.

Eads well understood now how serious this threat was. He and Dr. William Taussig, an executive of the bridge company, travelled to Washington, where they first talked with General Humphreys and, after getting nowhere, decided to call on Eads's old friend, now President Ulysses S. Grant. The president warmly welcomed them and, upon hearing their story, admitted knowing nothing of it and sent for General Belknap. Under Grant's questioning Belknap admitted that, yes, the bridge had been authorized as built by an act of Congress, but that he, as secretary of war, had the authority to remove obstructions to navigation. Grant's reply was short and to the point:

> You certainly cannot remove this structure on your own judgment. If Congress were to order its removal it would have to pay for it. It would hardly do that to save high smokestacks from being lowered when passing under the bridge. If your Keokuk friends feel grieved let them sue the bridge people for damages. I think, General, you had better drop the case.[20]

The steamboat fate was thus finally settled at the highest level of the nation, and from then on the railroads sped westward and linked the entire country together in a way impossible by boat. The railroad characterized U.S. engineering and commercial development more than any other means of transportation and probably more than any other innovation. It became the socially transforming network of the nineteenth century and its machine became the symbol of the move westward as well as the connection of town and country.

The Bridge Innovation

Although the bridge survived and the railroads prospered, St. Louis quickly lost its position as the most populous city in mid-America and even the bridge lost its financial standing. Railroads boycotted it, the bridge company went bankrupt, and the symbol of St. Louis's struggle passed into alien hands. In 1881, seven years after completion, the bridge received a rare gift: It was honored in a unique biography by C. M. Woodward, dedicated to the bridge itself, then called the "St. Louis Bridge."[21] Like Fulton's *Patent Application*, Francis's *Lowell Hydraulic Experiments*, and Henry's paper *The Galvanic Multiplier*, Woodward's *History of the St. Louis Bridge* is a primary cultural document in U.S. history because it presents the elements upon which innovation rests: the use of

Figure 8.2 The St. Louis Bridge

available theories, the connection of existing ideas and artifacts in un-precedented ways, and the competition over a design that begins a new di-rection for society. Above all, each document has at its center a single dominant figure: Fulton, Francis, Henry, and Eads. Partly, of course, these names are symbols of innovation as well as names of leading innovators.

Eads was so dominant a personality that those who wrote about the bridge, from Woodward on, have recognized that, as Eads's deputy W. Milnor Roberts wrote at the outset, "the Bridge, in its inception, in its plan, and in its noble battle against very fierce and extreme opposition,

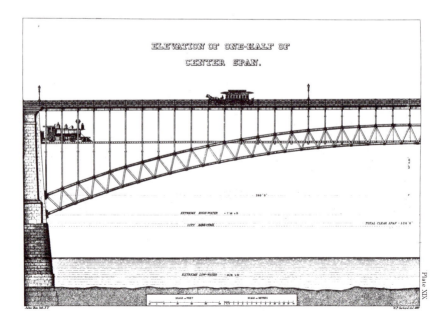

Figure 8.3 The St. Louis Bridge: Elevation of one-half of center span

is eminently yours."[22] Roberts was independently one of the most distinguished civil engineers of the late nineteenth century and his appraisal was echoed by others, including Dr. William Taussig, chairman of the bridge company's executive committee, who in a letter to Andrew Carnegie referred to Eads as the type of "Genius . . . [who] never permits itself to be fettered by theories, and I take it that his genius is more of a creative than of a theoretical turn. . . . He is a noble character, as well as a noble Engineer."[23]

This view of Eads led gradually to the practice, even as the bridge began to take shape in the early 1870s, of calling the structure the Eads Bridge. It then became the only major bridge in history to have as its official name that of its engineer. (The Cincinnati-Covington bridge was recently renamed the John A. Roebling bridge—so now there are two.) The most remarkable feature of the bridge is its form—a series of three arch spans—which so enthralled Eads that he argued for it against the prevailing attitudes of the engineering profession, the Corps of Engineers, and indirectly John A. Roebling, the leading U.S. bridge designer of the time. What was Eads's main idea?

The Arch of Strength and Beauty

Eads spoke of his 1867 design as "more commodious and attractive" than competing truss designs and he argued that the arch form best achieved "strength with durability and beauty with economy."[24] His central idea was both technical and aesthetic, the inseparable components of good bridge design. When the profession objected to Eads's 520-ft. span arches and proposed instead a bulky truss design with a maximum span of only 368 ft., Eads responded by invoking Telford's 600-ft. span cast-iron arch proposal of 1800:

> When we take into account that the limit of the elastic strength of cast iron in compression is only about 8000 pounds per square inch, and that in cast steel [for the Eads Bridge] it is at least seven or eight times greater, and consider the advance that has been made in the knowledge of bridge building since the days of Telford, it is safe to assert that the project of throwing a single arch of cast steel *two thousand feet in length* over the Mississippi, is less bold in design, and fully as practicable, as his cast iron arch of 600 ft. span.[25]

Eads based this remarkable defense and prediction upon the facts of metal arch behavior, which Telford had pioneered and in which the vertical gravity loads are converted into horizontal compressive forces

Eads Bridge

Model for Uniform Loads

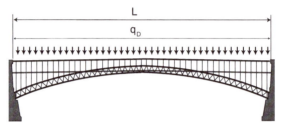

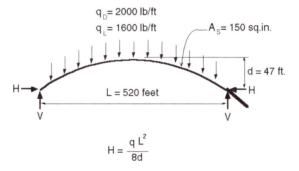

$q_D = 2000$ lb/ft

$q_L = 1600$ lb/ft

$A_s = 150$ sq.in.

$d = 47$ ft.

$L = 520$ feet

$$H = \frac{q\,L^2}{8d}$$

Model for equilibrium of uniform load on each of the four steel arches that make up the central span. Arch is assumed to be three hinged.

Calculations for Uniform Loads

Self-weight (uniform dead load)

$$H_D = \frac{2,000(520)^2}{8(47)} = 1,438,000 \text{ lbs.*}$$

*The axial compression increases from midspan to the supports by the factor $1/\cos\phi$ where ϕ is the slope angle of the arch. In this case the maximum increase is only 7 percent, which we neglect in this discussion. Eads provided more steel area near the supports to carry the increased forces there.

$$f_D = \frac{H_D}{A_S} = \frac{1,438,000}{2(75)} = 9,590 \text{ psi at midspan}$$

Full train load (uniform live load on whole span)

$$H_L = \frac{1,600(520)^2}{8(47)} = 1,151,000 \text{ lbs.}$$

$$f_L = \frac{H_D}{A_S} = \frac{1,151,000}{2(75)} = 7,670 \text{ psi at midspan}$$

Total stress

$$f_{TOTAL} = f_D + f_L = 9,590 + 7,670 = 17,260 \text{ psi at midspan}$$

This is a safe value because Eads considered the maximum allowable stress for steel to be 30,000 psi.

Model for Nonuniform Load

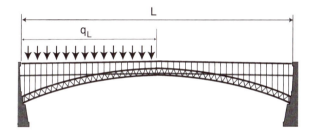

Bending in the Arch

Eads considered the possibility that heavy trains nonuniformly distributed on the bridge deck might lead to additional stresses in the arch. The following drawings illustrates the bending Eads anticipated in the half-loaded span:

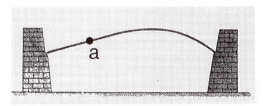

This bending, which is described by a "bending moment," M, introduces additional forces into the arch chords: compression (C) in the top chord and tension (T) in the bottom chord. These forces are inversely proportional to the distance between the chords, h. Eads chose $h = 12$ ft. a indicates the quarterspan.

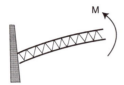

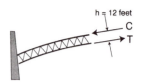

$$C = T = \frac{M}{h}$$

Calculations for Nonuniform Loads

For this case of half-loading, Eads calculated a bending moment at the quarterspan of 3,528 ft.-kips. (See Woodward, note 21.) So the additional force in one chord is

$$C = T = \frac{M}{h} = \frac{3,528}{12} = 294 \text{ kips}$$

Therefore, the compressive stress in the top chord would be

$$f_b = \frac{C}{A_S} = \frac{294,000}{75} = 3,920 \text{ psi*}$$

Had Eads chosen a closer spacing for the chords (such as $h = 3$ ft.), the forces and stresses would have changed as follows:

$$C = \frac{M}{h} = \frac{3,528}{3} = 1,176 \text{ kips}$$

Therefore, the compressive stress in the top chord would be:

$$f_b = \frac{C}{A_S} = \frac{1,176,000}{75} = 15,680 \text{ psi*}$$

With $h = 3$ ft., the stress increased by a factor of 4. Thus, Eads's design intelligently incorporates a large chord spacing, $h = 12$ ft., to minimize the influence of bending under nonuniform train loads. By splitting the arch into two widely spaced sections he is able to maintain the safety of the bridge without the use of added material (i.e., increased cost).

*At the quarterspan, the axial stress due to dead load is 9,740 psi and due to half-span live loads is 3,900 psi (assuming $A_S = 150$ sq. in.). Adding these values to the bending stress of 3,920 psi produces a total stress of 17,560 psi. So, in this case, the additional stress due to bending does not increase the arch stress much above the 17,260 psi found for the full uniformly distributed loads.

by giving the structure a flat, parabolic-like shape. For the Eads Bridge, the weight (q) of the fully loaded (dead load plus live load) bridge on each arch is about 3,600 lbs. for each ft. of bridge length. This means that on the main center span of length L = 520 ft. the total weight would be qL = 520 × 3,600 = 1,872,000 lbs.[26] The size of the horizontal force H at midspan acting on the arch cross-section is proportional to that weight and to the arch shape as defined by the ratio of the span, 520 ft., and the rise, d = 47 ft., which is the vertical distance between the arch supports and the midspan or crown of the arch. The formula for the horizontal force includes a constant 1/8 and gives

$$H = \frac{qL^2}{8d}$$

This is the same formula referred to earlier and used to find the horizontal tension force in the cable of a suspension bridge. There the term d represents the sag of the cables defined as the vertical distance between the tower tops and the low point of the cable at midspan.

For the Eads Bridge, the midspan compression would be

$$H = 1,872,999 \times \frac{520}{47} \times \frac{1}{8} = 2,590,000 \text{ lbs.}$$

Eads provided a cross-sectional area of cast steel, A = 75 sq. in. in each of the two tubes at midspan, so that the resulting compression stress[27]

$$f = \frac{2,590,000}{150} = 17,300 \text{ psi}$$

This is a safe value because Eads considered a safe maximum stress under all loads to be 30,000 psi.[28] The above result is for dead load (2,000 lbs. per ft.) and live load over the full span (1,600 lbs. per ft.). Such a load gives almost pure axial forces similar to those we found earlier in the boiler shell. In both cases, the form fits the load.

Engineers express this form principle by putting the design loads on a hanging string, observing the resulting curve of the string, and making the structure just that shape. For example, uniform loads oriented always perpendicular to the curve of the string produce a circular shape, just as the internal steam pressure requires a circular form, and uniform vertical loads produce the parabolic shape found closely approximate by suspension-bridge cables and the curve of a well-designed arch such as the Eads Bridge.[29]

When the live loads are not uniformly distributed over the entire span, that correct shape must change; in a flexible, cable-supported bridge it can usually do so. However, in a rigid, arch-supported bridge the shape cannot adapt to a different loading type and hence the arch bends, causing additional stresses. Eads provided for this bending by making the arches out of two tubes separated in elevation by 12 ft. The resistance to bending depends both on the material (two 75-sq.-in. tubes) and on its geometry (separation of 12 ft.). From Eads's calculations, the live-load bending creates at the quarterspan a compression force in the upper tube of 294 kips and a tension force in the lower one of the same 294 kips. The tension stress in the lower tube would be equal to the compression stress in the other tube:

$$f_L = \frac{294,000}{75} = 3,920 \text{ psi}$$

Had Eads spaced the same tubes at only 3 ft. then the bending would have caused a force four times as great, or 1,176 kips, for a compression stress in one tube of

$$f_L = \frac{1,176,000}{75} = 15,700 \text{ psi}$$

Therefore, it is clear that the geometry of the structure plays a crucial role in design, and by separating the two main parts of the arch Eads created a far safer design with no additional material.[30] These numerical results represent part of Eads's goal, which included as well a profitable utility and beautiful public structure: Efficiency, economy, and elegance combined in his mind to create a national landmark (which it now is).

Five Lines to the Pacific

Meanwhile, the railroads moved west with acceleration and without waiting for the St. Louis connection. In 1869 came the first transcontinental line connected in Promontory Point at the same time that Eads began construction of the caissons for his St. Louis connection.[31]

Following Promontory Point, other lines reached the Pacific including in 1883 the Northern Pacific, opened by President Chester A. Arthur on September 8 (just 3.5 months after he opened the Brooklyn Bridge).[32] The four California railroad men led by Stanford completed the Southern Pacific from New Orleans to the Pacific Coast.[33] Also completed in the 1880s was the Santa Fe Railroad, which went from Chicago through Kansas City, to Albuquerque, and into Los Angeles. The last line to the Pacific, the Great Northern Railway, reached Seattle in July of 1893.

Figure 8.4 Map of the railroads in the western United States

Within a generation after the Civil War five transcontinental roads had built through to the Pacific Coast. Their construction involved many unprecedented and complex problems of engineering and finance. The promoters, politicians, financiers, and contractors who planned, financed and built the lines lived in a lusty expansive America whose code of business ethics was often not high.[34]

The map shows these lines as they existed by 1893.

Railroad Competition in the East

Whereas Thomson had built a unified rail system from New York to Chicago by 1860, New York State lines were still by then separate companies. Cornelius Vanderbilt (1794–1877), having built a fortune in shipping and steamboating, turned to railroads late in life and, beginning in 1862, had by 1869 consolidated the lines between New York City and Buffalo into the New York Central system. His fortune rose thereby from

$11 million in 1862 to about $90 million by his death in 1877 and to $200 million by 1885 when his son William H. Vanderbilt (1821–1885) died. Both father and son ran the railroad for their own personal profit, and William made the widely quoted statement "the public be damned" during a newspaper interview on his private car near Chicago.[35]

A competition arose during the 1880s between the Pennsylvania and the Central, when the former threatened to build and operate a line parallel to the Central's in New York State. Vanderbilt retaliated by purchasing land running parallel to the Pennsylvania main line to Pittsburgh and even began construction, building tunnels and right-of-way. He received financial support from Andrew Carnegie (1837–1919), who wanted Vanderbilt to break the Pennsylvania's monopoly on rates into Pittsburgh.

The two railroad giants began cutting rates, which led to a drop in dividends, and that was too much for the nation's leading banker, John Pierpont Morgan (1837–1913), who entered the scene decisively. In July 1885 he invited the leaders of both railroads aboard his palatial yacht the *Corsair* and, while it cruised in a leisurely manner through the East River and the Long Island Sound, Morgan forged an agreement between the railroads that ended the war, sent rates and dividends back up, and netted Morgan by some estimates a fee of as much as $3 million.[36]

Regulation

Monopolies, excessive profits, and especially inflated rates caused a cry for regulation of the railroads after the Civil War. President Grant urged Congress in 1872 to study the problem, but little resulted although in the 1870s and 1880s individual states did attempt to legislate some control, with Illinois enacting strict regulation by 1873. Finally, in his State of the Union Address of 1883, President Arthur called on Congress to act in the public interest.[37]

John H. Reagan (1818–1905) got a bill passed in the House by 1878 and Shelby M. Cullom (1829–1914), former governor of Illinois, directed a study in 1885 of railroad abuses. In October of 1886, "the Supreme Court decided that a state could not regulate any rates on shipments passing beyond its own borders." Similar to the steamboat case in 1824, this ruling meant that the federal government would have to act and to take charge of this growing problem. Once again the tension between private enterprise and public welfare forced the Congress to debate and to legislate in a way not contemplated by the eighteenth-century framers of the Constitution. Engineering events were forcing the nation to create new political instruments.

On February 4, 1887, President Grover Cleveland (1837–1908) signed the Interstate Commerce Act created from the Senate's Collum bill and the House's Reagan bill.[38] One outcome was the creation of a five-person Interstate Commerce Commission with Judge Cooley as chairman to administer the act and to enforce its regulations. Although the commission served as a symbol of public concern, it could not easily control the activities of the financiers. For example:

> Much of the new concentration of railroad control had been accompanied by financial manipulations so unscrupulous that they would have excited the envy of the elder Gould, Drew, or Commodore Vanderbilt. Edward H. Harriman (1848–1909), George J. Gould, and James Stillman (1850–1918), head of the National City Bank of New York, were top men in a syndicate which financially ravished the Chicago and Alton between 1898 and 1905. The once well-run road was ruined as its capital structure was expanded from $34,000,000 to $114,000,000 with the insiders taking a private profit of at least $23,000,000.[39]

Finally a new political initiative was essential, and it came from a new and vigorous president, Theodore Roosevelt (1858–1919), who succeeded in pushing through legislation in 1903 that, among other things, made railroads as well as company officials liable to prosecution.[40] When Jay Gould (1836–1892) died, he left a fortune of at least $77 million and the *New York Times* referred to him as a pirate and a scourge.[41] The result, nevertheless, of nineteenth-century railroad building was the most impressive national rail system in the world and one that gave the country as a whole new opportunities for a rising standard of living.

The origin of this expansion of rail lines and of government control was the locomotive, with its power from reciprocating cylinders and its traction of driving wheels on rails. This prototypical nineteenth-century machine transformed the continent, its politics, and the art of the railroad age.

The Locomotive as Symbol

"There is nothing in the visible landscape—no tradition, no standard, no institution—capable of standing up to the forces of which the railroad is the symbol."[42] So Leo Marx typifies the attitude of nineteenth-century writers toward the most pervasive technological factor in U.S.

society at the time. It led Henry James to identify the locomotive as the machine entering the immense landscape of the United States and becoming "the germ of the most final of all generalizations" about the country.[43]

That entrance, according to Marx, first came symbolically in a wood near Concord, Massachusetts, on the morning of July 27, 1844, when Nathaniel Hawthorne's tranquil reflections on nature and the simple rural life were "interrupted by the long shriek, harsh . . . whistle of the locomotive . . . [which] brings the noisy world into the midst of our slumberous peace."[44] Marx goes on to observe that "it is difficult to think of a major American writer upon whom the image of the machine's sudden appearance in the landscape has not exercised its fascination."[45]

Not only did the railroad influence literature, but it also "took a firm grip on the imagination of many of the nation's leading artists [painters]." These painters include George Inness (1825–1894), especially in his c. 1855 *Lackawanna Valley*, which depicts a train in the countryside near Scranton, Pennsylvania; and, from twentieth-century painters, Edward Hopper (1882–1967) and Charles Sheeler, who was both a photographer and a painter.[46]

Not only did the railroad come through the garden with a shriek but it also dead-ended into the city. Here, in the largest cities of the country, the new machine had to be stopped, turned around, or changed. What the track was to the country the terminal represented for the city.

On September 1, 1869, the first stone was laid for a terminal in New York City on 42nd Street to serve Vanderbilt's growing railroads. The building was completed in 1871 and included semicircular wrought-iron arches 199 ft. in span, 94 ft. in rise, and covering a shed 652 ft. long running north from 42nd Street at 4th Avenue. By 1875 the tracks were depressed to 96th Street and by 1899 the terminal, having proven inadequate, was rebuilt in its present form and completed by 1901.[47]

All the while the city was growing at a remarkable rate. In 1890 its population reached 1,515,301 and by 1900 it had expanded and grown to 3,390,582. A major part of that extraordinary increase was due to the fact that on January 1, 1898, the city of New York (essentially Manhattan and the Bronx) incorporated Queens, Brooklyn, and Staten Island into its city limits. New York City reached its maturity in the 1880s and over the next 30 years created a remarkable set of new forms, largely thanks to the pressure from the railroads.[48]

All of this activity depended heavily on the one type of industry that had begun the industrial revolution in the first place: the industrial production of ferrous metal, which by the 1880s was characterized by the Scottish immigrant Andrew Carnegie and his gigantic steel empire.

9 Carnegie and the Climax of Steel

Carnegie the Epitome

No one personified the transformation of the United States more completely than Andrew Carnegie (1835–1919). His early life was a summary of the main engineering events of the time, and the contrast between his father and Andrew sharpens the dividing line between ancient and modern. The father, a hand weaver in Dunfermline 14 miles north of Edinburgh, followed a home working tradition essentially unchanged since medieval times; the son embraced mechanization and created the largest industry in the world. Even after power looms forced him out of business, Will Carnegie tried to sell his homemade wares by peddling them unsuccessfully door to door; Andrew by contrast at age 13 began work in a mechanized textile mill, immediately contributing to the immigrant family's income.

Thus the first business events in the family's U.S. life, after they arrived in Pittsburgh in 1848, were with textiles. But to get to Pittsburgh from their debarkation on July 15, 1848, in New York Harbor required a three-week inland journey up the Hudson, through the Erie Canal, west on Lake Erie, south on the Ohio Canal to the Ohio River, and finally upstream to their new home. Canals and textiles gave way quickly to the newest engineering marvel, the telegraph, and in 1849 Carnegie began work at the O'Reilly telegraph office. In 1851 he became a full-time operator and soon was regarded as the best in Pittsburgh.[1]

The following year Edgar Thomson appointed Thomas A. Scott (1823–1881) superintendent of the Pennsylvania Railroad's western division, and Scott, impressed by the 17-year-old Carnegie's telegraphy work, hired him as his secretary and personal telegrapher. Thus began Carnegie's association with the greatest of all midcentury businesses, Thomson's railroad. Quickly Carnegie learned about railroad operations, first through its telegraphic communications network and then more generally through its new style of management. Train movements had to be precisely controlled by communications and by well-disciplined workers. Accounts of sales and costs also had to be carefully recorded, and finally the complex railroad equipment had to be maintained, operated, and replaced in orderly and economical fashion. In his seven years working under Scott, Carnegie learned the business so thoroughly that when Scott became vice-president of the railroad, he appointed Carnegie superintendent of the same western division. All orders on the line between Altoona and Pittsburgh were now signed by A. C.

Carnegie did not participate in the original building as had Thomson, but he made many innovations in the operations and succeeded in raising profits so well that in 1865 Thomson offered him the position of general superintendent. The next step would be vice-president, but Carnegie declined. By 1865 he was ready for something even bigger. He had passed through all stages of the transformations of the country and now with the prospect of a postwar boom, he headed out into a new but closely related business.[2]

Carnegie: Bridges and Bonds

Carnegie could turn down Thomson's generous offer because by 1865 he was an independently wealthy man, but not due to his salary. He had earned a good living while learning about engineering as he worked directly with textile machinery, the telegraph, the steam locomotives, and derailments. He had also learned well the lessons of cost accounting and financial management. But at the same time Scott and Thomson had taught him the details of capitalist instruments: investments, dividends, and securities.

The year 1856 had been crucial for Carnegie. Scott lent him $600 to buy stock in a company, which almost immediately began paying him dividends. Stock prices rose; he could sell at a big profit, pay back his loans, and reinvest, and by 1863 he had an income from such dividends of more than $45,000 per year. He was rich at 28. He had learned the system so well that, because of his superior engineering training, he was later able to outmaneuver a formidable array of business rivals, including Cornelius Vanderbilt, Jay Gould, John D. Rockefeller, and J. P. Morgan. Although they knew capitalism as thoroughly as Carnegie, they had little firsthand feeling for the technology. Carnegie had the immense advantage of knowing how to create wealth rather than only how to manipulate it.

Figure 9.1 Andrew Carnegie (1835–1919)

In the same year that he began investment, he also began to think about another type of business: the building of bridges. John Piper, supervisor of bridge construction for the Pennsylvania Railroad, introduced Carnegie to iron bridges as replacements for the dangerously flammable wooden ones that locomotive sparks frequently ignited.

In 1862, Carnegie, encouraged by his growing wealth, organized a partnership with Piper and another bridge engineer to form the Keystone Bridge Company, which he considered to be "the parent of all [his] other works."[3] His one-fifth share cost him $1,250 (which he borrowed from a bank) and in 1863 it paid him $7,500 in dividends. Carnegie predicted a rapid explosion in bridge building after the Civil War, and with his independence in 1865 he could focus attention on securing the best contracts. He succeeded in getting the major and pioneering steel bridge contracts including the Eads Bridge arches and deck and the Brooklyn Bridge deck.

But for much of his time between 1865 and 1872, Carnegie pursued his skill as a capitalist and speculator, always related to his vast railroad experience: bridge building, telegraph line construction, sleeping-car manufacturing, the Union Pacific Railroad, and finally to finance all these ventures he became a supersalesman of U.S. bonds in Europe. Then at age 37, having mastered the railroad business and the art of capitalism, he shifted once again and for the last time; as he put it, "My preference was always for manufacturing. . . . I wished to make something tangible."[4]

Iron and Steel: Tangible and Tenacious

Iron and steel climaxed nineteenth-century engineering and society's transformation during that 100 years. In structures there were the Eads Bridge and the Brooklyn Bridge; in machines came the Francis turbine and the Baldwin locomotive, as well as networks of telegraphic copper activated by ferrous metal magnets and of industrial power transmitted by wheels and shafts to metal-working shops and metal-frame looms; and finally behind all those artifacts lay the basis for the entire industrial revolution: the metal itself, created by ancient processes that only gradually changed during the eighteenth century to allow for industrialization and to stimulate rapid mechanization.

Carnegie's experience with ferrous metal had begun in his early railroad days, and he saw especially the problems of iron rail deterioration under gradually increasing locomotive loads. Already in 1862 Thomson had decided to try steel rails for the Pennsylvania Railroad on the theory that reduced operating costs would justify the high investment, and by 1866 he considered the experiment a success. The steel rails lasted between five and eight times as long while only costing twice as much as iron rails.[5] At first the steel rails had to be gotten from England, where the new Bessemer process had just begun to produce steel in sufficient quantities to be affordable. By the end of the 1880s steel rails were costing less than iron ones.[6]

Although the introduction of cheap steel was a major innovation, it was industrialized iron—developed a century earlier—that made the industrial revolution. As with textiles, ironmaking is an ancient craft that changed only slowly until the eighteenth century. In both cases, the processes even of the late twentieth century are merely improved versions of practices that go back to prehistorical times. Indeed, the desire for clothes and tools distinguishes humans from animals, and it is therefore not surprising that these two products should be the defining features of the greatest social transformation in recorded human history, which symbolically begins in 1779 with the completion of Iron Bridge in the West Midlands of England.

From the cast-iron Iron Bridge to the Carnegie steel deck of Brooklyn Bridge, the engineering of ferrous metals changed radically in scale and incrementally in quality. For example, the tensile strength of eighteenth-century cast iron was about 10,000 lbs. per sq. in., whereas wrought iron then had about double that strength. By the late nineteenth century, cast iron could have 20,000 lbs. per sq. in. and wrought iron 40,000 lbs. per sq. in., but steel of the best engineering type had strengths of more than 200,000 lbs. per sq. in.[7] This immense difference along with a far smaller differential in cost accounts for the dominance of steel in the 1880s and the rise of that industry to its position as the world's largest.

Iron and steel transformed society because the tension strength is vastly greater than that of stone (which has barely 600 lbs. per sq. in.) and because ferrous metal is far harder than wood. Hardness means, technically, that heavy loads will indent iron far less than wood, a major reason that wooden rails quickly gave way to iron ones as locomotive weights increased in the 1830s. Moreover, steel traditionally had been differentiated from iron by its greater hardness, making it ideal for rails and machine tools.[8] The transformation in society occurred because these material properties were crucial for weapons and armor as well as for structures and machines.

Metal production requires two major engineering activities known as extraction metallurgy and physical metallurgy: how to get metal out of the ground and out of its ore and how to put that metal into a useful form, defined by hardness, strength, stiffness, and toughness (resistance to cracks). The first activity requires mining the ore and the fuel, usually coal, for heating it in furnaces to separate out the pure metal. The second activity, closely interrelated to the first, consists of playing with the pure metal, which is too soft for most uses, by treatments that include changing temperatures and adding other materials. This may sound unscientific, but it is a reasonable and historically correct description of how the making of steel evolved to its dominant position in the 1880s by traditional and empirical methods.[9] This tradition begins with pig iron, then wrought iron, and leads to steel; even the modifiers *pig* and *wrought* connote tradition rather than science.

Pig Iron

The ancient art of producing pig iron relied upon a blast furnace, consisting of a vertical fireproof vessel into the top of which are placed a charge of iron ore (containing 50–65 percent iron in the form of the iron oxides Fe_2O_3 or Fe_3O_4), coke or charcoal (carbon with sulfur impurities), and limestone (calcium carbonate), roughly in the proportions of 4, 2, and 1, respectively, by weight.[10]

As the coke in the blast-furnace charge descends, it burns rapidly in the hot blast air in front of the air inlet openings and reaches temperatures of hotter than 1,900 degrees C. The heating causes a chemical reaction that produces carbon monoxide, CO, which reduces (removes oxygen from) the iron oxides to give iron, Fe, and carbon dioxide, CO_2, or

$$Fe_2O_3 + 3CO \rightarrow 2Fe + 3CO_2$$

The primary function of the limestone in the furnace charge is to control the composition of the slag, the by-product material that contains most of the mineral impurities in the ore. In the upper part of the furnace, the limestone is calcined, or decomposed, by the reaction $CaCO_3 \rightarrow CaO + CO_2$. The proper ratio of basic constituents (CaO + MgO) to acid constituents ($SiO_2 + Al_2O_3$) must be maintained to obtain a fluid slag with good desulfurizing power. Maintaining a fluid slag is key to the operation of the blast furnace as a continuous process. The composition of the slag is controlled so the noniron materials are dissolved in it. Most of the sulfur impurity enters the furnace with the coke and forms FeS in an unwanted side reaction. The sulfur is removed from the iron by using the calcined limestone in the presence of carbon:

$$CaO + FeS + C \rightarrow CaS + Fe + CO$$

The slag (CaS plus other impurities containing aluminum, silicon, magnesium, etc.) is drawn off from an opening at the base of the furnace above that from which the heavier pig iron comes. The iron acquired the name *pig iron* because the molds were originally made of sand and formed in a manner that suggested a nursing litter of pigs.[11]

The resulting pig iron, when poured into molds made for a specific use, became the cast iron characteristic of the early industrial revolution—the Iron Bridge of 1779 being the first monument so built. The difficulty with cast iron, arising from the high carbon content, lay in its brittleness. It could not serve for tools or weapons and it had low tensile strength. Just as casting iron was an ancient process, so also was the means for overcoming its brittleness by reheating and hammering it. Pig iron contains about 4 percent carbon because of the formation of an iron carbide called cementite (Fe_3C) in the blast furnace by reactions like

$$3Fe_2O_3 + 11C \rightarrow 2Fe_3C + 9CO$$

Heating and hammering eventually removed most of the carbon to produce wrought iron by creating a coating of iron oxide, FeO, which combined with the cementite, Fe_3C, in the pig iron as follows:

$$Fe_3C + FeO \rightarrow 4Fe + CO$$

After long hours of working by the smith, this reaction gradually made almost pure iron. This wrought iron has good tensile strength and not the

Blast Furnace

Chemical Reactions

Coke (carbon) burns with oxygen in hot air (the blast)

$$C + O_2 \rightarrow CO_2 \tag{1}$$

$$CO_2 + C \rightarrow 2CO \tag{2}$$

to produce heat and carbon monoxide (CO), which is a strong reducing agent (removes oxygen) for iron ore, the most common being hematite, Fe_2O_3:

$$Fe_2O_3 + 3CO \rightarrow 2Fe + 3CO_2 \tag{3}$$

But coke has impurities such as sulfur that remove some of the valuable iron:

$$Fe + S \rightarrow FeS \tag{4}$$

Recovering this loss requires the addition of limestone (calcium carbonate, $CaCO_3$), which decomposes with heat

$$CaCO_3 \rightarrow CaO + CO_2 \tag{5}$$

and then removes the sulfur

$$CaO + FeS + C \rightarrow CaS + Fe + CO \tag{6}$$

The slag (CaS plus impurities containing aluminum, silicon, magnesium, etc.) floats above the iron and out of the furnace and the CO reduces more ore.

Formulation of the Problem

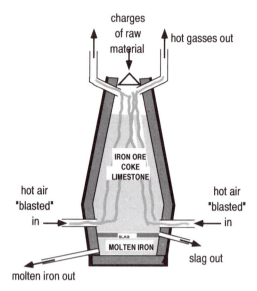

Collecting together equations (1) to (6) in terms of the charges of raw materials plus the hot-air blast (the *input* to the furnace), we can write (where Xs are unknowns)

$$X_1C + \ X_2S \ + X_3O_2 + X_4Fe_2O_3 + X_5CaCO_3$$

coke   sulfur   air   iron ore   limestone   (7)

Doing the same for iron, slag, and hot gases (the *output* from the furnace), we find

$$X_6Fe + X_7CaS + X_8CO_2$$

iron   slag   hot gas   (8)

We now need to compute the X values to find the quantities of charge needed to produce a certain quantity of iron. We note that intermediate products such as CO and CaO do not appear in the final equation because they cancel out: 3CO in (1) cancels 3CO in (6) and so on.

Solution

First we rewrite the input and output expressions as one equation (7) = (8):

$$X_1C + X_2S + X_3O_2 + X_4Fe_2O_3 + X_5CaCO_3 \rightarrow$$
$$X_6Fe + X_7CaS + X_8CO_2 \qquad\qquad\qquad (9)$$

Second we invoke the basic principle, already implied in each of equations (1) through (6), that atoms are conserved, meaning that the number of atoms of each element (C, O, S, Fe, and Ca) going into the furnace must equal the number coming out. The molecules (O_2, Fe_2O_3, etc.) are all transformed but the atoms are unchanged.

For example, $X_4Fe_2 = X_6Fe$, from which we see that $2X_4 = X_6$, and in a similar way we can get one equation for each element. This gives five equations with eight unknowns, which means we can arbitrarily pick the quantity of three elements and then solve for the remaining five.

Third, we shall begin by assuming the quantities of iron ore, coke, and sulfur to be $X_4 = 25$, $X_1 = 140$, and $X_2 = 10$. From these assumptions we find

$$2X_4 = X_6; \qquad X_6 = 50$$
$$X_2 = X_7; \qquad X_7 = 10$$
$$X_5 = X_7; \qquad X_5 = 10$$
$$X_1 + X_5 = X_8; \qquad X_8 = 150$$
$$2X_3 + 3X_4 + 3X_5 = 2X_8; \quad X_3 = 150 - \frac{3}{2}(25) - \frac{3}{2}(10) = 97.5$$

Thus the final expression would be

$$140C + 10S + 97.5\ O_2 + 25\ Fe_2O_3 + 10\ CaCO_3 \rightarrow$$
$$50\ Fe + 10\ CaS + 150\ CO_2 \qquad\qquad (10)$$

Material Quantities

The solution has so far given the quantities of molecules that balance equation (9) once we have specified three of the unknown quantities. A molecule is far too small a quantity to have practical or engineering significance. What we need are weights of materials. These come from the atomic weights as shown in the table below, which also gives the molecular weights by combining the individual atoms in a molecule. These are dimensionless numbers that give relative weight only.

Chemical engineers use the term *mole* to define the weight of a material that is numerically equal to its molecular weight . Therefore in equation (10), the 140 molecules of carbon is equivalent to 140 moles and weighs $140 \times 12 = 1,680$ lbs. The weights of each term are:

$25\ Fe_2O_3 = 25 \times 160 = $ *4,000 lbs.* of iron ore

$140\ C = 140 \times 12\ = 1680$

$10\ S = 10 \times 32\quad = 320$

$\left.\right\}$ *2,000 lbs.* of impure coke

$10\ CaCO_3 = 10 \times 100 = $ *1,000 lbs.* of limestone

We see that the charge of raw materials put into the blast furnace in this case is in the ratio of 4:2:1, a traditional ratio. This process would produce $50\ Fe = 50 \times 56 = 2{,}800\ lbs.$ of iron from the 4,000 lbs. of iron ore. The weights put in must be equal to the weights coming out by the *law of conservation of mass.*

Atom	Weight (lbs./lb.-mole)	Molecule	Weight (lbs./lb.-mole)
C	12	O_2	32
O	16	CO	28
Fe	56	Fe_2O_3	160
S	32	CO_2	44
Ca	40	FeS	88
		CaO	56
		CaS	72
		$CaCO_3$	100

brittleness of cast iron, but its laborious process made it too expensive for industrial use until the innovation of puddling by Henry Cort (1740–1800), who patented the process in 1784.[12]

Puddling consisted of hoeing the molten pig iron with iron rods inserted through small openings into a reverberatory furnace. Hot gases from the coke fire passed over the pig iron, combined with the carbon in the hot metal to form carbon monoxide, and were carried up the chimney, leaving almost pure iron. A puddler could produce about a ton of wrought iron in this fashion compared to the 100 or 200 lbs. a man could make by hammering. Puddling replaced hammering and wrought iron became "the key to the Industrial Revolution," until 1855, when Henry Bessemer (1813–1898) entered the iron world and started its transformation to a steel age.[13]

Bessemer Steel

Most steel contains greater than 98 percent iron. However, it also contains carbon, which, if present in sufficient amounts (up to about 2 percent), gives steel a property unmatched by any of the metals available to the ancient world. This property is the ability to become extremely hard if cooled very quickly from a high enough temperature, as by being immersed in water or some other liquid.

Pig iron may contain 3–4.5 percent carbon, 0.15–2.5 percent manganese, 0.2 percent sulfur, 0.025–2.5 percent phosphorus, and 0.5–4.0 percent silicon. In refining pig iron to make steel, these elements must be either removed almost entirely or reduced drastically in amount. It was the Bessemer process that first made this possible economically and on a large scale.

Bessemer's father was a skilled machinist, metal worker, and inventor. With this background, the son began, with little formal education, to work with machines, with metal, and with new ideas already at age 17. After a series of inventions, Bessemer developed a new machine to produce bronze powders far cheaper than anything on the market. He proceeded to set up a manufacturing plant, went into production, and made enough money to permit him to focus on further inventions.[14] He had turned an invention into an innovation, and in later years he attributed his success to the fact that "I had no fixed idea derived from long established practice to control and bias my mind, and did not suffer from the too general belief that whatever is, is right."[15]

This description somewhat generalized could apply equally well to Fulton—painter, canal designer, bridge designer, submarine inventor, and steamboat innovator; to Morse, who left a major career in painting to explore the telegraph; to Francis, who went easily from railroads, to canals, to turbines, and to structures; and indeed to all the people we have featured up to this point. This attitude would apply to Thomas Edison and even to others beyond our present scope. Indeed, it is not too much to argue that the narrowing of specialization in the twentieth century has often worked against innovation. As Hector Berlioz is reported to have said in response to the question of why Saint-Saëns, thought to have been the most knowledgeable and gifted musician of his day, never became a great composer, "he lacked inexperience."[16]

By 1851 Bessemer held about 25 patents, and shortly thereafter he became intrigued by a new problem, the firing of shells from cannons. On November 30, 1853, the Russian Black Sea fleet complete with new naval guns destroyed a Turkish fleet in the harbor of Sinope.[17] The event had two major results: one was the Crimean War whereas the other, obscure to most histories but at least as influential, stimulated Henry Bessemer to think about projectiles and armored vessels.[18] By early 1855, Bessemer had recognized that the principal difficulty lay in casting a cannon strong enough to withstand the forces needed to eject his newly designed, heavy projectiles. He saw clearly that he must study metallurgy and this quickly led him to building a furnace for making better iron.

Bessemer Process, 1856

Converting Iron to Steel

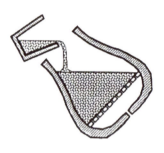

1. Molten pig iron is poured into the pear-shaped Bessemer converter.

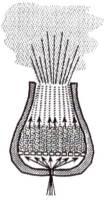

2. Air at 20+ psi is blown thorough the molten pig iron, oxidizing the impurities and producing a great burst of flame lasting 10–12 min.

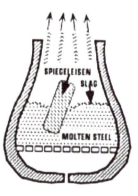

SPIEGELEISEN

SLAG

MOLTEN STEEL

3. Excess oxygen is removed by reacting with a small amount of spiegeleisen (pig iron containing about 20% manganese).

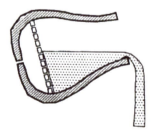

4. Molten steel is poured into ingot molds prior to being rolled into shapes.

Figure 9.2 Henry Bessemer (1813–1898)

In early 1855, there occurred an accidental event that startled Bessemer and set off a new train of thought:

> Bessemer happened to notice several pieces of pig iron that lay at one side of the bath, unmelted by the intense heat of the furnace. In order to increase the combustion, he turned on more air. Half an hour later he opened the furnace door to check on the progress of the heat and observed that the two pieces of pig iron still remained unfused. With the intention of pushing them into the molten metal, he poked them with an iron bar. To his surprise, he saw that they had become thin shells of decarburized iron. This phenomenon led him to understand that atmospheric air alone [without burning a fuel] was capable of decarburizing pig iron completely and of converting it into malleable iron.[19]

After more than a year of painstaking study, building new vessels, and increasing the air blown into the iron, he asked one of Britain's most distinguished engineers, George Rennie, to see a demonstration of his new process. Rennie was so impressed he urged Bessemer to present it at the forthcoming meeting of the Mechanical Section of the British Association for the Advancement of Science at Cheltenham. There, on August 11, 1856, Bessemer delivered his paper "The Manufacture of Malleable Iron and Steel Without Fuel," probably the most memorable paper in the history of metal.[20]

His progress created a storm of interest, but when others tried to apply his methods under license, they failed. For almost two years Bessemer retreated to his laboratory to find out why. It was during this period that science entered to help him identify the culprit, phosphorous, which so weakened his product as to make it useless. Bessemer hired a distinguished chemist as well as several other scientists. Together with the inventor they located the difficulty and Bessemer then sought and found phosphorous-free iron ore. By 1858 he had built a factory at Sheffield where he could produce steel to sell for 6 to 9 pounds sterling per ton against the prevailing price of 60 pounds per ton by older processes. By 1860 he began making a profit; 14 years later the profits totalled 81 times the original capital investment. Steel had arrived, and soon there would arrive at Sheffield another engineer intent on bringing steel to the United States: Alexander Lyman Holley (1832–1882).

Alexander Lyman Holley

Ten years before Andrew Carnegie turned to steel manufacturing, the U.S. engineer Alexander Holley arrived in Europe and made his way to Sheffield to see the Bessemer works firsthand. He was deeply impressed there by "the converters erupting with their roaring cascades of dazzling sparks."[21] The year before, Holley had published a massive volume entitled *American and European Railway Practice in the Economical Generation of Steam*, in which he included a section on Bessemer steel and one on the advantages of steel, which, he wrote, "will gradually come into exclusive service, not only increasing the safety . . . but promoting the economy."[22]

During a second trip to Sheffield in 1863, Holley negotiated an agreement with Bessemer for U.S. rights to his process, and upon his return Holley designed in Troy the first U.S. Bessemer steel plant. Design meant building, testing, tearing apart, rebuilding, and trial after

Figure 9.3 Alexander Lyman Holley (1832–1882)

trial until on April 27, 1865, his Bessemer converter finally produced good-quality steel.

A quarter of a century later in a lecture dedicated to Holley, James Dredge, the distinguished editor of *Engineering*, described that event in Troy as "the beginning of the Bessemer steel industry in the United States. . . . I ascribe all the credit to Alexander Lyman Holley. It was Holley who adapted the English details of the process to fit American requirements, and who initiated that system of rapid production, combined with a high standard of excellence, which aroused at first incredulity, and afterwards the admiration of European manufacturers."[23] Who was this man unknown to U.S. history?

Born in 1832 in Lakeville, Connecticut, Holley showed an early interest in machinery and especially locomotives. His father planned on his entering Yale, but the 17-year-old looked with horror on the prospect of four years in Latin and Greek. The elder Holley relented when he discovered that Brown University was planning a new curriculum emphasizing science and applications. So in the fall of 1850 Alexander Holley began studies in Providence.[24]

Following graduation, Holley was hired by America's most famous steam-engine builder, George Corliss of Providence. There he spent more than a year building a locomotive to his own design. It worked, but not well, and it convinced Corliss to give up on mobile engines. But he did not give up on Holley; Corliss wanted him to stay in the firm but Holley was obsessed by locomotives and left to find work only six months later in Jersey City starting in September 1855.

He worked not only on locomotive designs but also on drafting; his elegantly colored drawings of locomotives caught the eye of Zerah Colburn, editor of the *Railroad Advocate*. Holley had begun to write articles for various magazines while with Corliss, and in April 1856 he left the Jersey City locomotive works to become part owner and editor of Colburn's journal, which got renamed in August *Holley's Railroad Advocate*. Between 1856 and 1861, Holley wrote articles for the *Advocate*, *The American Engineer*, *The American Railway Review*, and *Van Nostrand's Eclectic Engineering Magazine*. From late 1858, he wrote editorials and more than 200 articles for the newly formed *New York Times* (founded 1851) on engineering of every type.[25] During this time he co-authored with Colburn a lavishly illustrated book on European and American railways.[26]

His work on steel began in 1862 when, after many financial trials, he secured a commission from Edwin Stevens, of the Stevens family from Hoboken, New Jersey, who wanted a full report on European armor to be used in his design for an ironclad boat for the Civil War. Holley and his father, who had been governor of Connecticut, left for Europe in August 1862.

The elder Holley was intent on countering the British sympathies for the South in the Civil War, whereas his son focused primarily on studying British armaments. Holley wrote incessantly—articles for the *New York Times*, the *Atlantic Monthly*, the *National Almanac* of 1863—and eventually he published a major treatise, *Ordnance and Armor*, with sections on steelmaking. This 1865 volume of some 900 pages with 493 illustrations became a source book of practice in the steel industry.[27] It served to launch Holley's career and turn him finally away from locomotives and the railroad and toward the emerging business of designing and building steel plants in the United States.

As the direct result of his trip to England, Holley began in 1863 to promote steelmaking in the United States. His first success was at Troy, New York. Then he moved to Pennsylvania near Harrisburg, and finally, in his largest enterprise of all, Holley designed and built the J. Edgar Thomson Steel Works about 12 miles from Pittsburgh.

It was here that Holley and Carnegie came together in 1872 to create the nation's largest steel mill, and the new firm of Carnegie, McCandless and Co. formed to produce Bessemer steel. The organization had the business genius of Carnegie and, in Captain Bill Jones, a mill superintendent who understood the ideas of both the other two plus the means of working with factory laborers. The plant began to produce steel in 1875, and by 1884 it was turning out 500,000 tons per year—four times that of its closest rival in Pittsburgh.[28] Moreover, the productivity of the new mills was staggering by European standards. By 1876 there were 27 Bessemer converters in the United States producing about 526,000 tons of steel, compared to 110 converters in England producing only 700,000 tons.[29] Thus the average yearly U.S. productivity was 20,000 tons per converter whereas in England it was only 6,400 tons.

Innovation in Steel

Holley had designed ten Bessemer converter plants prior to the Thomson works, Jones had managed large steelworks before coming to Pittsburgh, and Carnegie had substantial business experience prior to 1872. Thus the Thomson works were not novel; the steelmaking processes were worked out earlier in England and even at Troy and other sites in the United States.

This history is similar to that of Lowell, Massachusetts, where a new large-scale industrial organization grew up on the basis of novel works done elsewhere. In both cases, the pioneering inventions took place in England and men of exceptional engineering talent studied there to bring back to the United States as much technical knowledge as they could. Lowell memorized the textile mill machinery in Birmingham; Holley detailed the steelmaking at Sheffield. Both then set about to design larger, more fully integrated industrial processes aimed at the potentially larger markets in the United States.

Indeed, Fulton also took directly from England in the form of Watt's engine and Beaufoy's tests results, just as the first U.S. railways took English locomotives and rail-line designs. These American engineers had an international outlook: They knew that engineering abroad was an essential part of their education but they also knew that engineering exists in a cultural context. Therefore, they could study abroad but they had to know the political and economic conditions at home if they hoped to succeed in building new systems of transportation or new industrial organizations.

These men were individual geniuses: Fulton, Morse, Francis, Lowell, Thomson, Carnegie, Eads, Corliss, and Holley. But they arrived in their maturity at a time and place of expansion and of a developed engineering context. Without experienced iron workers, machinists, and manufacturers, they could have accomplished little; there needed to be a strong technological infrastructure. Finally, as each worked on large, complex new designs, each found the need to know more about how things worked and that need led to closer study of structures, machines, networks, and processes. Such studies led them into the unfamiliar world of stresses, heat, circuits, and materials and gradually brought scientists and scientific analysis into the practice of engineering. At the time the J. Edgar Thomson mills were completed and profitable, Bill Jones could say to Carnegie that "chemistry might ruin the goddamn industry."[30] Yet Carnegie, without directly disputing Jones, hired a German chemist and thence "nine tenths of all the uncertainties of pig iron making were dispelled under the burning sun of chemical knowledge."[31]

Since Carnegie's time the role of science in engineering has always been subjected to those conflicting views: those of the production superintendent and those of the technologically literate entrepreneur. It is thus that the ideas of applied science, social process, and individual genius combine to explain innovation in the modern technological world. By the death of Holley in 1882, each idea had found expression in his articulate, sometimes passionate writings; in his detailed and successful designs; and in his broad, international outlook.

The Building of an Industry

When Carnegie decided to enter the steel business, he had not only the right kind of technological experience but also some original ideas on manufacturing. He needed four basic components to put together from scratch his new business. He needed capital, a site for his mill, a group of lieutenants, and a management plan. It was Carnegie's genius to recognize these essentials and to integrate them into a single enterprise whose time had come.

Carnegie secured *capital* from a number of local people, including a respected Pittsburgh merchant and family friend, David McCandless. Carnegie capitalized on the merchant's strong business reputation by naming his new firm Carnegie, McCandless and Co. He also got his former patrons Tom Scott and J. Edgar Thomson to invest; it did not hurt that he named the new plant for the president of his largest potential client.[32]

The *site* was at Braddock's field (where it cost less than land nearer Pittsburgh), next to two railroads plus the Monongahela River. Carnegie did not want to be at the mercy of only one railroad even if it were the one run by the mill's namesake. Then he needed to get the best *people*, and for that he found Holley and, through him, Jones.[33] Finally came Carnegie's *management plan,* which consisted of a continual accounting procedure to track closely week by week all costs, of a goal to run the mill at full capacity all the time, and of a clear line of accountable supervisors.

Carnegie reserved for himself two essential functions: watchdog over the entire operation and salesman to ensure full production. His ultimate goal was to cut prices so he could gain control of the market, and then, as he put it, "profits will take care of themselves." Moreover, he continually plowed profits back into improvements in the plant, just as had Thomson with the railroad. [34] He rewarded his key people with partnerships but kept their salaries low. Captain Jones, however, refused partnership and requested instead "a hell of a big salary." Carnegie made that request his one exception and gave Jones the unheard-of sum of $25,000 a year, equal to the salary of the president of the United States.[35]

Between 1875 and 1883 Carnegie transformed the steel industry, made millions of dollars, and even thought about retiring to pursue his goal of cultural enrichment and philanthropy. But events changed his mind and he kept on. He would finally sell at the start of the new century, but that story, full of conflict and conquest, belongs to another part of the transformation of U.S. society, which began with the railroad and steel as well as with an entirely new industry that grew out of the telegraph and the steam engine: the industry characterized by a man as eccentric and successful in his own way as Andrew Carnegie was in his.

10 Edison and the Network for Light

Carnegie and Edison

From the 1840s until the 1880s the railroads and the steel industry dominated U.S. engineering, economics, and politics and engaged the attention of the general public and the federal government. Following the 1876 centennial, however, a new force entered big business. Just as steel is popularly associated with one man, Andrew Carnegie, so the new force of electricity came to be the realm of one unique character in U.S. history: Thomas Edison (1847–1931).

Like Carnegie, Edison was born without wealth and began his full-time working life as a child west of the Alleghenies. He, too, gained local fame as a brilliant telegrapher. This early technical talent and self-sufficiency led Edison by the age of 23 to a financial independence in 1870, just as those same traits had gained personal wealth for Carnegie by 1863.

By 1872 Carnegie settled on the emerging steel industry as his full-time activity, while Edison by 1876 began to focus on directing a research and development laboratory. Both men exhibited substantial skill at promotion and both imagined large-scale operations aimed at applying inventions to profitable industrial applications: Carnegie with the processes of steelmaking and Edison with networks of electrical power. Both built great systems and by the early 1880s both had revolutionized engineering and set in motion the pattern for urban, industrialized life in the twentieth century.

Both entrepreneurs left the scenes of their successes to pursue other activities later in life: Carnegie, his philanthropy, and Edison, his laboratory. The similarities are striking but so are the differences, and these differences reflect as much the differences between processes and networks as they do the differences between the two extraordinary characters who pioneered such systems in the late nineteenth century.

The process tended toward concentration of production whereas the network spread outward into large regions. The process requires the bringing together of natural resources initially spread over a wide area; the network sends out artificial products usually made in centalized locations. Networks tend toward public utilities whereas processes tend toward private monopolies. Edison invented small devices to serve a multitude of small consumers and Carnegie built huge converters to make products (rails, for example) for immense industries (railroads). Edison was the inventor without peer and Carnegie the entrepreneur extraordinaire. Edison had a plan for public relations but he was more a private, reclusive person, whereas Carnegie, with substantial technical talent, was a man of the world, skilled in an industry of conversion. The ultimate product of that industry was money, and no one amassed more of it in our history than the little Scottish imigrant.

Edison and the Lightbulb

Thomas Alva Edison, born in Milan, Ohio, in 1847, began his career as a telegraph operator in the Midwest. In 1868 he went to Boston, bought a set of Michael Faraday's works, and after going through them began to invent electrical devices. In 1869 he moved to New York City where, in 1870, he invented the glass-domed stock ticker, which was really a special type of telegraph that could print 185 characters a minute. It was his first major invention and with the $40,000 that the Gold and Silver Telegraph Company paid him for it, he established himself in Newark as a manufacturer of such tickers and other telegraphic equipment.[1]

In the centennial year of 1876 Edison moved to Menlo Park, New Jersey, to establish a research laboratory, and in 1878 he began to think in detail about a system of electrical lighting. As had occurred with Watt and his separate condenser, Fulton and his boat, Stephenson and his boiler tubes, and Francis and his turbine, Edison thought about his lamp as part of a working design and not as an isolated device to be perfected on its own. Unlike those predecessors, however, Edison visualized a completely new system rather than a new object. He saw right away that the lamp needed to be thought of as part of a distribution line and a central power plant. He even gave an interview with the *New York Sun* on October 20, 1878, in which he described his system before he had even worked out its design.[2]

Figure 10.1 Thomas Alva Edison (1847–1931)

Edison worked closely with a number of associates in Menlo Park including Francis R. Upton, a mathematical physicist trained at Bowdoin and Princeton, throughout late 1878, and by October 1879 he had designed his first practical lamp. Edison recognized the need for a high-resistance lamp because he understood the implications of the two basic electric laws: Joule's law and Ohm's law. As one of his associates put it at about this time:

> Edison's ideas were far in advance and his able mathematician, Francis R. Upton, translated them into figures. . . . Upton at first did not understand [the need for a high-resistance lamp], but Edison in his own way, using matches to illustrate his points, soon convinced him.[3]

Following that observation, the associate proceeds to give Upton's numerical proof of Edison's insight, which appears here in the figures.

The major factor in the lamp design was the need for power in the lamp to produce light. The electric power P is measured in watts and comes from Joule's law:

$$P = I^2 R$$

where I is current (in amperes, or amps for short) and R is resistance (in ohms). Edison decided that he needed 100 watts of power to have an electric lamp that would compete with gas lamps. In a system of many lamps connected in parallel, the total current required is the sum of the currents drawn by each lamp, and the voltage across each lamp, V (in volts), is the same, given by Ohm's law:

$$V = IR$$

Thus, Edison could choose the lamp current and resistance to obtain 100 watts of power in Joule's law; the required power-plant generator voltage was then determined by Ohm's law. For example, with $R = 100$ ohms and $I = 1$ amp, $V = 1$ amp $\times$ 100 ohms $= 100$ volts.

Figure 10.2 Francis Upton (upper left) with Edison Group

Edison's Lighting System

Scientific Basis: Upton's Analysis

Joule's Law: $P = VI = I^2R$

Transmission line resistance

$$R_S = \rho\,\frac{L}{A}$$

L = length of line
A = cross-section of wire
ρ = resistivity of material (a constant)

Power loss in a transmission line

$$P_S = R_S\,I^2 = \frac{\rho L I^2}{A}; \quad A = \frac{\rho L I^2}{P_S}$$

Examples

For a given ρ and P_S:

(1) $I = 10$ amps; requires a line cross-section

$$A = \frac{100\rho L}{P_S}$$

(2) $I = 1$ amp; requires a line cross-section

$$A = \frac{\rho L}{P_S}$$

Conclusion: The transmission line for a lamp working on 1 amp requires a cross-sectional area 100 times less than that required for a line serving a lamp that uses 10 amps.

Assumption: The percent loss of total power through line losses is the same for each case.

Upton's Calculation of Line Losses

Ohm's law: $V = IR$
Joule's Law: $P = VI = I^2R$
Edison's criterion: $P_L = 100$ watts

Low-resistance lamps

Resistance $R_L = 1$ ohm
Curent required $I = 10$ amps
Lamp power $P_L = 10^2(1) = 100$ watts

A (line) $= 0.01$ sq. in., $L = 1,000$ ft., $\rho = 0.67 \times 10^{-6}$
ohm-in. (copper wire)

R (line) $= 0.8$ ohms

Power losses in the lines, $P_S = 10^2(0.8) = 80$ watts

Power supplied $= P_L + P_S = 100 + 80 = 180$ watts

Line losses are $80/180 = 4/9$ of total power

High-resistance lamps

Resistance $R_L = 100$ ohms
Curent required $I = 1$ amps
Lamp power $P_L = 1^2(100) = 100$ watts

A (line) $= 0.0001$ sq. in., $L = 1,000$ ft., $\rho = 0.67 \times 10^{-6}$
ohm-in. (copper wire)

R (line) $= 80$ ohms

Power losses in the lines, $P_S = 1^2(80) = 80$ watts

Power supplied $= P_L + P_S = 100 + 80 = 180$ watts

Line losses are $80/180 = 4/9$ of total power

Conclusion: For the same loss of power, the system with the low-resistance lamps (1 ohm) requires 100 times more transmission line cross-section [A(line) $= 0.01$ sq. in.] than the cross-section [A(line) $= 0.0001$ sq. in.] needed for high-resistance lamps (100 ohms).

Social [Economic] Basis for Lighting

Having a high-resistance filament lamp, Edison needed to design a system with low line losses by also reducing

$$R_S = \rho \frac{L_S}{A_S}.$$

Thus, he

1. Used a good conductor, copper, with low $\rho = 0.67 \times 10^{-6}$ ohm-in.
2. Kept the power plant close to the lamps to keep L_S low (of the order of 5,000 ft.).
3. Realized the need for relatively high A_S.

But high area of copper meant high cost so that low I_S and short lines (low L_S) became crucial. Edison's economic analysis showed the overriding need for keeping the copper conductor cost in line:

	100 ohms 1 amp 100-watt bulb	25 ohms 2 amps 100-watt bulb
Power plant	$86,680	$86,680
Conductors	$57,000	$228,000
Miscellaneous	$7,000	$7,000
	$150,680	$321,680

The choices of R and I were made by Edison on the basis of a fundamental economic implication of Joule's law for the entire power network, which was that distribution lines in the network also have resistance, so Joule's law applies, except that the appropriate current and resistance values must be used. The line resistance, denoted by R_s, is given by

$$R_S = \rho \frac{L}{A}$$

where ρ is a material constant (called resistivity), L is the length of the wires, and A is the wire cross-sectional area. In a system with a given material, ρL will be fixed and only A can be varied. As A is increased (to reduce R_s and hence power loss in the lines), the cost of the wires (copper costs for Edison) would increase. Because Edison estimated that the cost of the copper conductors was nearly 40 percent of the entire system capital cost (including power plant, lines, and meters), a major increase in A would easily make electric lights more expensive than gas lighting.[4] Edison recognized that the alternative to a larger A is to arrange for lower-line current I, thereby requiring a higher resistance filament to maintain the normal 100-watt power requirement for the lamps.

Edison was thinking about a network of about 1,000 lamps. For 100-watt, 100-ohm lamps (as above), the total line current would be 1,000 amps. If the same system were used with 100-watt, 25-ohm lamps, the total current doubles and the power loss in the lines increases by a factor of four. To reduce the line loss to the same as before by using bigger wires would require an increase of A by a factor of four; this is also the factor by which material costs would increase, and the economy of the design would be severely compromised. Using Edison's calculations, we would find that the increase in copper wire section could have increased his cost estimate for the entire system by more than 100 percent.

Edison and the Power Plant

Thus Edison realized that he needed a high-resistance lamp filament to keep the current low and hence the line losses low as well. The search for the practical high-resistance filament took most of 1879. Once he had found a successful filament, Edison turned to the design of a complete system and in September 1882, he opened the Pearl Street Station in New York City of the Edison Electric Illuminating Company. The age of central power plants and incandescent light had commenced. As Edison wrote of his approach:

> It was not only necessary that the lamps should give light and the dynamos generate current, but the lamps must be adapted to the current of the dynamos, and the dynamos must be constructed to give the character of current required by the lamps, and likewise all parts of the system must be constructed with reference to all other parts, since, in one sense, all the parts form one machine, and the connections between the parts being electrical instead of mechanical. Like any other machine the failure of one part to cooperate properly with the other part disorganizes the whole and renders it inoperative for the purposes intended.
>
> The problem then that I undertook to solve was stated generally, the production of the multifarious apparatus, methods and devices, each adapted for use with every other, and all forming a comprehensive system.[5]

The approach was typical of other leading engineers such as Watt, Fulton, Stephenson, and Morse. They all tried to visualize the complete work even as they sought to improve single elements within it.

Driving north on the New Jersey Turnpike in the 1990s and into the technological density of suburban Newark, we come upon the Bayway Refinery on the left and Public Service's Linden Power Plant on the right. What we cannot see is the intimate subterranean connections: refined oil flowing from Bayway to Linden, powering the plant, whose electricity returns to Bayway to power the refinery. Yet when the first electric-power generating station began a few miles to the northeast, its goal was to eliminate oil products for lighting, not to encourage them.

Edison began electrical power generation with the goal of competing with gas lamp illumination in New York City. His thought was systematic and he visualized a network of wires carrying the power to the lamps with as little loss as possible. He sought to prevent the conversion of electrical power into heat (resistance in the lines) until the final destination (the lamps). The idea of this network, which Edison pioneered, was to deliver something with as little change as possible. He carried this feature of changelessness too far by resisting conversion to alternating current, but the idea of the network did not change even with that more flexible system of current.

The Pearl Street Plant

The world's first central electricity-generating power plant for incandescent lighting not only illustrates Edison's systematic approach and attention to details but also shows his attention to competitive economics and to the need for Wall Street backing. Many engineers were working on lighting systems after the introduction of the practical direct-current generator, first by the Belgian Zénobe Gramme in 1870 and the German Werner Siemens in 1872.[6] At first most systems were arc lights arranged in series, so the resistance of each lamp had to be low to avoid the need for excessive power. If one lamp failed, the entire system failed like a chain with a weak link. Other people in the 1870s recognized the desirability of incandescent lamps.

Electric current crossing a gap between two electrodes causes a luminous discharge to create arc lamps, whereas electric current passing through a filament to cause it to emit light when heated creates incandescence. Charles F. Brush (1849–1929) had pioneered the use of arc lighting in the late 1870s and succeeded in having the first central electric generating station built in 1879 in San Francisco. Like Edison he developed a system including his own dynamo (direct current generator); it proved practical and was widely used in the 1880s.[7]

Likewise, other inventors worked on incandescent lamps at this same time. Nevertheless Edison remains the single most important pioneer because of his recognition that innovation requires three components: applied science, a social process, and individual genius. The Pearl Street plant illustrates his understanding of innovation probably better than any other event of his remarkable career.

Edison's collaboration with Upton characterizes his advanced ideas on the application of scientific principles to the practical ends, for example, of low line loses and the high-resistance filament. By scientific, we mean the mathematical formulation of physical principles that allow the engineer to predict accurately the performance of a technological work. In the case of the Pearl Street design, Upton's analysis (see figures) provided Edison with that formulation and guided the design for all parts of the system.

At Pearl Street Edison bought six steam engines, each rated at 125 Hp with a maximum power of 200 Hp. These each drove one jumbo dynamo designed by Edison and named after the elephant brought to the United States by P. T. Barnum. These dynamos each had to light 1,200 lamps with 0.75 amps and 110 volts arranged in parallel circuits.[8] The power required to light all these lamps would be 99 kilowatts, or 1.34(99) = 133 Hp where 1 kilowatt is equivalent to 1.34 Hp.

By keeping the line losses low, Edison required very little extra power (about 9 Hp for a 1,000-ft. feeder line) to provide the full capacity to the 1,200 lamps. The filament to each lamp had, however, to be of high resistance (about 100 ohms) so the line current could be kept reasonably low to limit line losses. Edison's initial lamps had carbonized paper filaments to produce the high resistance.[9]

Of great significance was the need to measure the power supplied. Edison, like Francis a generation earlier, had to devise a means for measuring the flow of his power-producing quantity. For Francis that had been water flow (Q), and for Edison that had to be electricity flow I, or current. In Francis's case, power came then by multiplying Q by the water fall or potential energy; for Edison power derived from the product of I and the electric potential, V, voltage. In both cases the potentials were reasonably constant (H for water and V for Edison's parallel circuits).

Edison took advantage of the well-known electrochemical effect (electroplating), which established that a current of 1 amp flowing for 1 hour would deposit 1,213 milligrams of zinc on a receptor plate. His problem, like Francis's, was to measure flow without disrupting performance.

Edison achieved this by designing a divided or shunt circuit on one side of which was a shunt of very small resistance and on the other side was a compensating coil of great resistance in series with the measuring cell. The line current to be measured, therefore, was split between the coil and the shunt with 0.1 percent going to the coil and cell and 99.9 percent going through the shunt. This small current then slowly deposited the zinc, whose weight was measured periodically by a meter man and the client charged accordingly.[10]

These charges then lead to the second component of innovation, the social process by which electric power would prove to be economically and politically acceptable to society. Edison made an economical analysis to compare his estimated costs of building and operating a central power station to the costs associated with gas lighting. This analysis guided his design by showing the central importance of the copper feeder wires. The table shows Edison's estimate.[11] Although later research indicates that his estimate was far in error, it nevertheless played a crucial role in his design thinking and illustrates that his motivation was a competitive product, not a string of accredited patents.[12]

Even more important economically was Edison's friendship with the lawyer Grosvenor P. Lowrey. It was Lowrey who recommended Upton to Edison and it was Lowrey who connected the inventor to the Wall Street financiers. In 1866 Lowrey had become general council to the Western Union Telegraph Company; there he met Edison in connection with lawsuits over the telegraph.[13] Lowrey saw Edison's potential and actually suggested that he go into electric lighting. But Lowrey also did much more. He secured capital for Edison through his contacts with the firm of Drexel, Morgan and Company. Lowrey's office was on the third floor of the Drexel Building, an ornate, white marble "skyscraper" of seven stories, on the corner of Broad and Wall Streets.[14] The year 1879 saw not only the Edison lamp but also the ascendancy of a young banker whose father had died that year. That year began the independent career of J. P. Morgan (1837–1913), whose influence on U.S. industry—railroads, electricity, and steel—would parallel that of the engineers Thomson, Edison, and Carnegie.

Lowrey's service as financial advisor was indispensable to Edison the innovator. Edison needed money to keep his growing Menlo Park laboratory at work and especially to push it toward commercial profits. In 1880 Lowrey arranged an extravagant private-train visit to Menlo Park for New York politicians to see a spectacular lighting display aimed at overcoming the lobbying efforts of the gaslight industry. Thus Lowrey's help was both political and economic, and Edison responded with his power network of 1882.

From purely scientific and social standpoints, this event would not have been successful had it not served another purpose. The Pearl Street plant was part of a system even larger than its Wall Street network; it was part of a system of Edison enterprises that included a series of power plants, the Edison Electric Light Company, the Edison Machine Works to build dynamos, the Edison Electric Tube Company to make underground conductors, and the Edison Lamp Works to manufacture incandescent lamps. All of these were in place *before* Pearl Street opened.[15] Pearl Street was not a commercial success if taken alone, but it served the larger purpose of symbolizing the potential for electric lighting everywhere. By the late 1880s companies or utilities owned by Edison and his backers existed all over the United States and as far abroad as Milan and Berlin.[16] The symbol from Wall Street was beginning to light the nation.

But even more than the plant on Pearl Street, Edison himself became the primary symbol for electric lighting. He had created his labora-

Pearl Street Power Plant

Power

Thomas Edison bought six steam engines to power his 1882 Pearl Street electrical generating station in New York City. Each engine produced a maximum of 200 Hp and the following values for *PLAN*:

$P = 73$ psi (estimated average pressure)

$L = 16$ in. or 1.33 ft.

$A = (11\tfrac{3}{16})^2 \pi/4 = 98$ sq. in.

$N = 700$ power strokes per min.

These gave

$$P_E = \frac{PLAN}{33,000} = \frac{73(1.33)98(700)}{33,000} = 200 \text{ Hp}$$

Each steam engine powered one electric dynamo designed to light 1,200 lamps. For each lamp,

$V = 110$ volts

$I = 0.75$ amps per lamp

According to Joule's law, power equals volts × amps, therefore,

$P_L = VI$ (number of lamps) $= 110(0.75)1,200 = 99$ kilowatts (kw)

$P_L = 99(1.34) = 133$ Hp, where 1 kw = 1.34 Hp

The steam engines produce more power than necessary to compensate for the inefficiency of the dynamos and power losses in the feeder lines that connected the station to the street lamps.

Power Losses

The power losses in the feeder lines depend, first, upon the *line resistance* where

$L = 1,000$ ft. (estimated feeder length)

$\rho = 0.67 \times 10^{-6}$ ohm in. (resistivity of copper)

$A = 1$ sq. in. (estimated area of feeder lines) according to the relationship used by Edison for calculating resistance

$$R_S = \rho \frac{L}{A} = 0.67 \times 10^{-6} \frac{12,000}{1} = 0.008 \text{ ohms}$$

Power losses in the feeder lines depend, second, upon the *current* in the line:

$I_S = 0.75(1,200) = 900$ amps

Combining Joule's law ($P = VI$) and Ohm's law ($V = IR$), we find $P = I^2R$. For power losses in the line, therefore,

$$P_S = I_S^2 R_S = (900)^2 \, 0.008 = 6.48 \, \text{kw} = 8.7 \, \text{Hp}$$

Each dynamo has enough power to overcome the line losses and supply the lamps:

$$P_g = P_L + P_S = 133 + 8.7 = 142 \, Hp$$

At maximum steam-engine power, the system efficiency (power in the lamps divided by power supplied) is

$$\frac{133}{200} = 66.5\%$$

Incandescent Lamps

Carbon filament lamp
Filament size:

$A = 0.4 \times 10^{-4}$ sq. in. (estimated)

$L = 3.13$ in.

$\rho = 1.6 \times 10^{-3}$ ohm.-in. (carbon)

$$R = \rho \frac{L}{A} = 1.6 \times 10^{-3} \, \frac{3.13}{0.4 \times 10^{-4}} = 125 \, ohms$$

From Ohm's law,

$$R = \frac{V}{I} = \frac{110}{0.75} = 147 \, ohms$$

(probably the filament area was smaller than 0.4×10^{-4} sq. in.)

Electric Light Meter

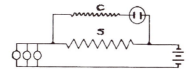

Compensating coil resistance, $C = 9.99 \, \text{ohms}$; $I_C = \dfrac{1}{1,000} I$; where I is the total current in the lamps.

Shunt resistance, $S = 0.01 \, \text{ohms}$; $I_S = \dfrac{999}{1,000} I$

1 amp-hour deposits 1,213 milligrams of zinc
Meter constant $= 1,213/1,000 = 1.213$

A reading of 1.213 (measured weight of deposited zinc) means 1 amp-hour of electricity was used. Because the voltage was roughly constant, this reading could be interpreted as kilowatt hours.

tory, he had directed the lamp project, and he had formed companies to sell products. He had a flair for publicity and the tireless energy of both a focused researcher and an exuberant promoter. People called him the wizard of Menlo Park and they marveled at his prodigious production of inventions; during the design stages for Pearl Street, Edison applied for 256 patents (1880–1882) all under his own name.[17]

It is obvious from the detailed scholarly studies on Edison that he could not alone have perfected that massive output. He had highly talented associates such as Upton; Francis Jehl; John Kreusi, his Swiss master machinist; and another master craftsman, Charles Batchelor. Once again it is the great-man theory versus the teamwork idea about innovation. Taking firsthand testimony along with the documentation of Edison's notebooks, it seems correct to affirm both that Edison was an individual genius, a true wizard, and that at the same time he needed a team of highly talented, even inventive people in order to create so much.

The symbol of Edison—the overwhelming leader in patents, in lighting, and in research—contributed to the raising of capital and to the defense of patent rights. Edison's prestige was a commodity and it was largely based on merit, even if some colleagues thought they deserved more public credit (Jehl) and some competitors openly questioned his greatness (Tesla).[18] But after a century of argument, Edison's place in history is secure, even if after the huge success in the 1880s his judgment did seem to be flawed in one significant event: the war of the currents between direct and alternating.

The Transformation to Alternating Current

A distinguished historian, Isaiah Berlin, published a book in 1953 entitled *The Hedgehog and the Fox* in which he noted that "the Fox knows many things but the hedgehog knows one big thing." Berlin identified Dante, Plato, Dostoyevsky, and Ibsen as hedgehogs, and Thomas Hughes in his prize-winning book *Networks of Power: Electrification in Western Society, 1880–1930* observes that Thomas Edison belongs in that group.[19]

Hughes describes Edison as someone who conceptualized and solved problems "associated with the growth of system . . . [and] his historical peers were other inventor-entrepreneurs, such men as Robert Fulton, Samuel Morse, and Cyrus Hall McCormick [who invented and then manufactured the reaper]." But the hedgehog who knows one big thing can sometimes fail to see another big thing growing up nearby. And that other big thing challenged Edison's big thing frontally and ultimately successfully. The two things were two types of electrical current: direct current and alternating current.

Edison had chosen to use direct current in 1878 largely because the dynamo was then well developed. Like Watt building on Newcomen and Fulton using a Watt engine, Edison took as many working components as he could find. When designing his first central power plant, Edison was careful to pick a densely built up region in lower Manhattan where the transmission lines could be short but the number of customers high.

This choice was a little like Brunel's decision to use a wide-gauge line between the closely spaced, densely populated cities from London to Bristol. There, a restricted network was economical if inflexible. In the same general way, the direct current system of Edison was a restricted network because of the power losses in the line, $P_s = I_s^2 R_s$. If there were 1,000 amps in the system with 100 volts (to light 1,000 100-watt lamps), the line losses would be equal to 1,000,000 R_s. Because the line resistance increases with the length of the wires in the network ($R_s = \rho L/A$), to keep power losses low and efficiency high required that L/A be kept small.

Viewed another way, the length could be increased only if A were increased in the same proportion, but increases in the wire size involved prohibitive material costs, and the practical consequence was that the distribution lines in Edison's system could not be longer than 1 or 2 miles. Far from permitting large central power stations serving wide geographic areas, Edison's system depended on neighborhood generating plants. In fact, the Pearl Street plant was followed soon after by about 60 similar electric power plants in the city of New York alone. Such intensity of construction made sense for a densely populated city but was hopeless for the Tennessee Valley Authority, where one would need to fill the entire rural valley with power stations to serve widely separated small towns and farms.

Alternating Current

The key idea for achieving efficient long-distance power transmission is the use of high voltages. Combining Joule's law and Ohm's law, power $P = I^2 R = I(IR) = VI$ (the product of current and voltage), so a high-voltage lamp will require less current to provide the same amount of power. Less current needed for the lamps means less power loss in the distribution lines. Engineers can create high-voltage direct current (DC) power generators by connecting several dynamos in series. At the "consumer end" of the system, however, safety considerations require that the lamp voltage be no higher than about 400 volts, and lower voltages are preferable for making long-lasting lightbulbs. Edison could increase the voltage to reduce line losses but he could not then easily reduce it for safe use in offices and homes.

The solution to this dilemma was alternating current (AC), a sinusoidally varying current. As early as the 1830s Michael Faraday in England and Joseph Henry in Princeton had explored the electromagnetic

Alternating Current

The Transformer

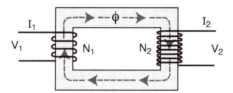

In the ideal transformer the flux, ϕ, is constant. The magnetic potential due to the primary current I_1 is equal and opposite to that due to the secondary current I_2:

$$I_1 N_1 = I_2 N_2$$

or the relationship between the current of the primary winding I_1 and that of the secondary winding I_2 can be written

$$\frac{I_1}{I_2} = \frac{N_2}{N_1}$$

Power supplied by the primary winding $V_1 I_1$ will be nearly equal to that induced into the secondary winding $V_2 I_2$, hence we may take

$$V_1 I_1 = V_2 I_2$$

Therefore,

$$\frac{I_1}{I_2} = \frac{V_2}{V_1} \quad \text{and} \quad \frac{V_2}{V_1} = \frac{N_2}{N_1}$$

effect known as mutual inductance.[20] This phenomenon arises from the interaction of two current-carrying wire coils in close proximity, whose magnetic fields occupy overlapping regions of space. A change in the amount of current in the circuit containing one coil induces a voltage change in the circuit containing the second coil. The magnetic fields of two coils can be designed to be closely confined in a common region by winding both coils around the same core of iron (or other magnetizable material). Such a device, known as a transformer, permits voltage to be greatly increased at the power plant, transmitted in high-voltage lines over long distances, and then greatly decreased at substations to be safely used in living quarters.

Joseph Henry first devised a transformer in 1838 while a physics professor at Princeton, and Elihu Thomson in 1879 gave perhaps the first public demonstration of stepping voltage up or down.

The Transformer

In the transformer (see above figure) the number of turns in the primary winding N_1 and the secondary winding N_2 determines the relationships between current and voltage in the windings:

$$\frac{I_1}{I_2} = \frac{V_2}{V_1} = \frac{N_2}{N_1}$$

Therefore, if the ratio of turns of wire, N_2/N_1, is 100, the transformed current will be from $I_1 = 1,000$ amps to $I_2 = 10$ amps and the transformed voltage from $V_1 = 100$ volts to $V_2 = 10,000$ volts. The power loss in the second high-voltage line will be $P_s = 10^2 \times R_s = 100 R_s$ instead of 1 million R_s as before. The power is thus sent over high-tension (high-voltage) lines with low current, giving rise to negligible line losses. The transformation only takes place for alternating current, where changes in the current through a first coil cause changes in the transformer's magnetic field which in turn induces a voltage in the second coil.

In a circuit with current I alternating at 60 cycles per sec., I increases from zero to a maximum in one direction, back to zero, then to a maximum in the other direction, and finally back to zero, thus completing one cycle in one-sixtieth of a sec. The current increases rapidly at first (the slope of the curve is large) until at the maximum its increase is zero (the curve is horizontal or has zero slope at the top). It is this increase that induces a voltage V_2 into the second circuit. It seemed obvious to a few pioneers in the mid-1880s that the future belonged to alternating current because of the need to transmit power over extensive networks. It was not so, however, to Thomas Edison.

The Battle of the Currents

Edison, having correctly begun with one system—easier and safer—could not bring himself to take the newer one seriously. George Westinghouse of Pittsburgh, the inventor of the air brake for locomotives, did take alternating current seriously and began in 1885 to manufacture equipment for alternating current operating with voltages of 1,000 volts and more.[21]

Edison fought back by trying to discredit alternating current in the most spectacular and grisly ways. In 1888 he arranged a series of public experiments where "hapless dogs and cats were nudged out onto wired sheets of tin and electrocuted by the high voltage of alternating current." Then, as the tour de force, he proposed a new device to replace the death penalty by hanging: the electric chair, using, of course, alternating current and Westinghouse equipment. This sorry episode in Edison's career culminated on August 6, 1890, when a condemned murderer, William Kemmler, was put to death in Auburn Prison, New York, by a Westinghouse machine. Early in his career Edison had used publicity effectively in promoting his ideas about power, but by 1890 such tactics used negatively did nothing to counteract acceptance of alternating current.

Nikola Tesla

The engineering figure who stands out in the historic battle of the currents is Nikola Tesla, "a towering mystical genius out of Yugoslavia" who had worked for Edison between 1884 and 1886 but who went with Westinghouse in 1888 just as the battle was joined. Tesla developed the alternating current motor, which permitted electrical power sent over long lines to be converted to mechanical power at distances remote from the central generating plant.[22] Westinghouse's system was exhibited at the Columbian Exhibition in Chicago of 1893 and it stimulated Henry Adams (grandson and great-grandson of presidents of the United States and a leading U.S. writer) to think about those great dynamos (strictly speaking, alternators). "Probably this was the first time since historians existed, that any of them had sat down helpless before a mechanical sequence. . . .

Chicago was the first expression of American thought as a unit; one must start there."[23]

Tesla was born in Yugoslavia in 1856.[24] His father was a Greek Orthodox priest and his mother the descendant of a long line of inventors. He studied engineering at Graz Polytechnic in Austria and finished his education at Prague University at the age of 23. At Graz, he saw one of his professors running a generator as a DC motor for a class demonstration. The brushes and commutator were sparking. It was obvious to him that they would soon wear out. Furthermore, the sparking represented a loss of energy and efficiency. He knew there had to be a better way. Five years later, as he was walking with a friend in the Budapest city park, the better way came to him in a flash of insight. He saw a simple way to produce a rotating magnetic field through polyphase currents. This brilliant discovery would be the heart of his AC motor. The commutator and brushes were not needed.[25]

Five more years were to pass from this moment of insight to his applications for five patents relating to his AC motors and the power production and distribution systems that went with them. In Budapest, he worked for a short time for the state telephone system and, in 1882, moved to Paris, where he worked for the French Edison Electric Light Company. In 1884 he sailed to the United States and arrived in New York with 4 cents to his name. He found employment in the Edison Machine Works, being introduced to the famous Edison through one of the shortest letters of recommendation ever written. It read, "I know two great men and you are one of them; the other is this young man." [26]

Edison developed a strong respect for the young inventor, but a salary dispute and the AC-DC question would lead Tesla to quit Edison in less than one year. In short, Edison would have nothing to do with Tesla's AC generation, transmission, and motor system. Tesla quickly found employment with a small group of promoters looking for a practical arc light for street and factory illumination. In one year, Tesla developed the desired lamp and several patents, but he gained little compensation from it.[27]

Eventually, Tesla was introduced to A. K. Brown, vice-president of the Western Union Telegraph Company, who put up the money to get Tesla started in his own company, the Tesla Electric Company (33–35 South Fifth Avenue, now West Broadway). Here Tesla produced three complete systems of AC machinery using single-phase, two-phase, and three-phase currents. For each, he produced the dynamo for generating the currents, motors for producing power, transformers for raising and lowering voltages, and a variety of devices for automatically controlling the machinery. He patented five of his inventions, but many more would follow.[28]

In 1888 he was invited to give a lecture to the four-year-old American Institute of Electrical Engineers (AIEE) at Columbia University in New York City. Through such lectures and his many patents, he quickly became recognized as the father of the whole AC power system and one of the most outstanding inventors in electricity.[29]

Tesla had no specific thoughts to commercialize at this time. It was George Westinghouse, a man of vision, who approached the 32-year-old inventor and offered to buy Tesla's 40-odd patents for $60,000 plus a royalty at $2.50 per Hp. Later, when Westinghouse was in dire financial straits in the 1890s, Tesla, in one of the greatest gestures of either superb kindness or utter stupidity, agreed to accept a cash settlement of $216,000 in lieu of the royalties. It has been estimated that this cost Tesla around $12 million at the time and possibly $50 million over his lifetime.[30] Furthermore, Tesla agreed to come to Pittsburgh for a year at a "high salary" to help Westinghouse commercialize the AC system. Just as when he worked with the Edison Company, Tesla had difficulty working with Westinghouse engineers.

Other inventors also were working on the AC induction motor in the late 1880s. Although many have claimed priority to the invention, it is clear that only two men, Nikola Tesla and Galileo Ferraris, need to be considered seriously for the honor.[31] Ferraris had apparently thought about the problem of producing a rotating magnetic field in 1885, but made no further progress. Tesla made his discovery in 1882 and within a few months developed a complete system, which included all that he eventually patented. In fact, Ferraris concluded that the rotating magnetic field could never be used to develop a motor.

At Westinghouse, Tesla worked on the design and construction of a streetcar induction motor. But the Tesla motor ran at a constant speed and had a low starting torque; neither characteristic was good for streetcar motors. Also, it ran best with low-frequency currents and not at the high frequency (133 Hertz) then in use. Furthermore, the Tesla motor used two-phase power, but all AC generators at the time were single-phase. Because the Tesla motor did not live up to its expectations and because of financial difficulties within the company, Westinghouse stopped work on it in December 1890. Tesla left Westinghouse in late 1889 to get back to what he did best: work alone in his own laboratory and invent.

11 The Centennial Revolutions, 1876–1883

The Centennial Exhibition

The history of the United States, as presented to most students, leaves out the Centennial Exhibition in Philadelphia of 1876. [1] As an event it seems to historians to be insignificant compared to the social issues that followed the Civil War. Yet the fair provides an archaeological storehouse for those engineering developments in the United States that were transforming society in the nineteenth century. In 1976, the director of the Smithsonian stated that

> in 1876, the grandest of all celebrations of America's Centennial was the one called "The United States International Exhibition." Staged in Philadelphia, this exhibition quickly acquired the nickname it has kept ever since—it was simply, "the Centennial." The Centennial was, as one journalist wrote, "the most stupendous and successful competitive exhibition the world ever saw. [2]

The Centennial Exhibition opened on May 10, 1876, in Philadelphia when President U. S. Grant and Emperor Dom Pedro of Brazil turned on the 39-ft. high Corliss engine that both powered Machinery Hall at the fair and provided its most enduring symbol. [3] At the same time, in New York, John Roebling's 276-ft. high stone towers stood complete in the East River in preparation for the 1,595-ft. span that would be started later. [4]

Neither the stone bridge tower nor the steel steam engine served any direct commercial or social use in 1876; both, however, stood for such use in the most powerful possible ways. That is to say, they both stood as icons to the emergence of the United States as the world's most powerful

Figure 11.1   Corliss engine at the 1876 Centennial Exhibition in
   Philadelphia

nation.[5] They were both self-conscious monuments that, consistent with American pragmatism, were intended to be directly useful as well as symbols of power and achievement.

These two works of pure technology represented the most visually obvious of the two primary objects of modern technology: the structure and the machine—the fixed, permanent object and the moving, transient object—that have completely reordered the world over the past 200 years.

The tower and the engine for 1876 stood at the dividing point in that two centuries of development: George Corliss's engine is an enlargement of James Watt's steam engine, and John Roebling's Brooklyn Bridge is a scaled up version of Thomas Telford's 1826 Menai Straits Bridge. Each object of 1876 represented the culmination of a vast nineteenth-century reorientation of British and U.S. society away from rural, isolated communities and toward nations closely knit by new networks that began to focus on industrialized and fast-growing cities. Therefore, each centennial monument stands for large-scale technology, which encouraged the building of high-density cities and in turn led aesthetically sensitive observers to contemplate the promise and the portent of a new type of civilization.

By 1883 engineering in the United States had become a dominant force whose growth we have characterized by four leading ideas: structure, machine, network, and process. Having traced these ideas through the steamboat, the water-powered textile mill, the railroad, the telegraph, steelmaking, and electric power, we can now summarize the ideas by making further comparisons and connections.

The Four Great Ideas of Engineering

The figures here illustrate the parallel nature of structures and machines through a comparison of their salient features: load, form, speed, result, and the connecting formula. In each case a load produces a result by means of the object's form, except that the machine requires motion or speed whereas the structure does not. Another way to view these relationships is to see their transforming action. The vertical bridge weight becomes a primarily horizontal cable force thanks to the form expressed as a ratio of span to sag (L/d). Nature exerts a downward pull (qL) while the structure resists by its horizontal tension (H).

For the machine, the natural pressure of the steam acting on the piston head (PA) becomes power (Hp) thanks to the motion of that piston (LN). These prime movers use a natural force to create a motion that results in mechanical power, just as the large-scale structure uses a natural force to create a form that results in structural resistance.

These two complimentary objects of engineering, expressing the ideas of structure and machine, require the complementary systems of networks and processes, which the two figures here illustrate, using the electric power network and the blast furnace process. Each system has a resource put into it, an agent that activates it, a valuable result for which the system is designed, and waste products that the designer seeks to minimize.

The network uses a natural resource (e.g., fuel, water, wind) activated by a generator and a closed circuit (through Ohm's law) to light a set of bulbs with the electric power ($I_L^2 R_L$) remaining after system line losses ($I_S^2 R_S$). In the same general way, the process takes a natural resource (iron ore), activated by burning carbon and getting carbon monoxide, which removes the oxygen from the ore to create iron along with the waste gas carbon dioxide.

Again, another way to view these relationships is to see them as transforming systems. The mechanical power of a rotating engine (Hp)

Structures and Machines

Symbols for Structures (cable or arch bridges)

H = horizontal thrust in the arch or cable at midspan (lbs.)
q = load (lbs. per ft.)
L = span (ft.)
d = rise in the arch or sag in the cable at midspan (ft.)

Symbols for Machines (reciprocating engine)

Hp = horsepower (33,000 ft.-lbs./min.)
P = pressure (lbs./sq. in.)
L = piston stroke (ft.)
A = piston head area (sq. in.)
N = strokes (number per min.)

	Structure	Machine
Load	qL	P
Form	L/d	LA
Speed	—	N
Result	H	Hp
Formula	$\dfrac{qL^2}{8d}$	$\dfrac{PLAN}{33,000}$

becomes electrical power in the generator transmitted through the circuit to end up emitting light through the power consumed by incandescent lamps. Similarly the useless compound of iron ore becomes the invaluable material iron through the highly reactive carbon monoxide, which itself is the product of a transformation of carbon through oxidization (burning).

From these examples, we see also how these four ideas connect. The production of electric power mostly comes from the burning of carbon (fossil) fuels in the chemical process of combustion, whereas the mechanical power of the steam engine transformed the result of that chemical process into the electrical power of the generator, which was made of iron and steel. The making of iron and steel is, too, a prerequisite for all mod-

Transforming Nature by Structures

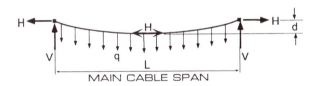

MAIN CABLE SPAN

$$qL \quad \times \quad \frac{L}{d} \quad \rightarrow \quad H$$

bridge   bridge   cable
weight   form   force

Transforming Nature by Machines

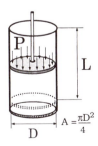

$$PA \quad \times \quad LN \quad \rightarrow \quad Hp$$

steam   piston   horsepower
force   motion

ern, large-scale structures and especially for the suspension bridge cables designed on the basis of the cable force (H). In these ways, the central ideas of structure, machine, network, and process weave together the complex patterns of engineering in the modern world.

The Tower and the Engine

These complex patterns of interconnection are expressed by and yet transcend formulas and thus require broader perspectives. The Roebling towers and the Corliss engine, like all works of modern technology,

---

Networks and Processes

Symbols for Networks (electrical power)

V = voltage (volts)
I = current (amps)
R = resistance (ohms)
Subscripts: L = lamp filament
s = system conductor lines

Symbols for Processes (iron smelting)

Fe = iron (chemical symbol for iron atom)
O = oxygen (chemical symbol for oxygen atom)
C = carbon (chemical symbol for carbon atom)
Subscripts: 2,3,etc. = number of atoms of an element in one molecule

	Networks	Processes
Resource	$\mathrm{Hp} \rightarrow VI$	Fe_2O_3
Agent	$V = IR$ (Ohm's law)	$2C + O_2 \rightarrow 2CO$
Result	$I_L^2 R_L$	Fe
Waste	$I_S^2 R_S$	CO_2
Formulas	$P_{in} = VI$ (Joule's law)	$2C + O_2 \rightarrow 2CO$
	$P_{out} = I_L^2 R_L + I_S^2 R_S$	$Fe_2O_3 + 3CO \rightarrow 2Fe + 3CO_2$

have a scientific basis, a social context, and symbolic power. These are the three aspects of technology we have sought to define in this text. As we have seen, to understand a work in scientific terms, we can describe it through simplified, but essentially correct, numerical calculations. For a structure, efficiency means the use of the least amount of material consistent with its required strength; for a machine, efficiency means the use of the least amount of energy consistent with its required power.

Efficiency as the scientific measure is, of course, only one aspect of technology, because saving in materials and energy must always be balanced by saving in cost and maintenance. These monetary measures are

Transforming Nature by Networks

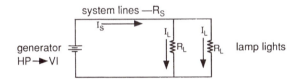

$$\begin{array}{ccccc} \text{Hp} & \rightarrow & 2I_L^2 R_L & + & I_2^S R_S \\ \text{mechanical} & & \text{lamp} & & \text{system line} \\ \text{power} & & \text{lights} & & \text{losses} \end{array}$$

This network power comes from a machine.

Transforming Nature by Processes

$$\begin{array}{ccccccc} Fe_2O_3 & + & 3CO & \rightarrow & 2Fe & + & 3CO_2 \\ \text{useless} & & \text{dangerous} & & \text{valuable} & & \text{waste} \\ \text{material} & & \text{gas} & & \text{material} & & \text{gas} \end{array}$$

For a structure H is the force in the steel cable carried by this material, Fe.

not scientific, but social; they depend upon time and place and characterize the political and economic world in the same way that efficiency characterizes the physical and chemical world. Cost for a structure, an object of public works, is prototypically political, and nothing reveals late nineteenth-century urban politics more clearly than the Brooklyn Bridge.[6] Cost for a machine, on the other hand, usually relates to private profit rather than public benefit; but at the same time, just as public works need to be economical, so do the products of private industry need to benefit and not harm public life.

Ultimately, of course, technology must be also evaluated in terms that elude quantitative measurement but that account for those fundamental aspects of life that we describe by such words as *beautiful*, *ethical*, and *spiritual*. The Brooklyn Bridge is one of those objects of pure technology that has so impressed both the individual artist and the general public that it has become not only a social benefit to transportation, but also an elegant park for the city resident, as well as an icon of even greater power than its cableless towers of 1876 revealed. For many artists, such as the poet Hart Crane and the painter Joseph Stella, the bridge has become a spiritual event, a celebration of the human spirit comparable with the greatest works of art.[7] It is, in fact, recognized to be a great work of art in its own right.

In the same way, the Corliss engine struck many leading artists of its day as a supreme aesthetic object that even inspired a spiritual awe and wonder.[8] Here again, we can find the marked difference between structures and machines: mass and energy, public benefit and private profit, and now a permanent symbol as opposed to a transient one. But the motion of the machine creates a different and more immediately striking response than the repose of the large-scale structure. The aesthetic of the machine, like that of the dance, the play, and the symphony, is transient but no less impressive than that of the structure, whose aesthetic is more like that of architecture, sculpture, painting, and photography.

The structure and the machine are not built to be isolated monuments, but to be parts of large technological systems. The Brooklyn Bridge was intended as a primary link in the intra-urban transportation network of metropolitan New York. The Corliss engine formed the heart of the fully mechanized Centennial Exhibition, to which its power spread out through long rotating shafts, and then in its second life that great engine powered an entire industrial community just south of Chicago. The isolated tower and the exhibition engine became, by 1883, the centerpieces of the most striking urban experiments in the United States: the connection between the first and third largest cities (New York and Brooklyn, then an independent city); and the United States' first fully planned industrial town, Pullman, Illinois, an outgrowth of America's soon to be second largest city, Chicago.[9]

These two objects, therefore, serve within two sorts of systems: the transportation network and the industrial process. As with electrical systems, the network is a static system comparable to the static structures that help form it; the process, as in steelmaking, is a dynamic system comparable to the dynamic machines that help power it. In Pullman, Illinois, the Corliss engine provided power to convert raw materials and prefabricated parts into a finished railroad car, whereas in New York, the Brooklyn Bridge provided the means for carriages and tramways to carry people and goods, essentially unchanged, from one part of the city to another.

Soon after the centennial the engine would be used as the center of a new network carrying electric power throughout the city, just as other bridges had already, since 1830, been helping to form a network of railroads connecting cities. In a similar way, steam engines had already entered the factory to produce chemical products as well as to power mechanical assemblies. In a general way, therefore, we have characterized modern engineering as composed of objects and systems, each of which has two complementary types. Structures and machines are the objects that correspond roughly to networks and processes, which are the systems.

John A. Roebling's Design: Scientific Aspect

John Roebling (1806–1869) based the design of the Brooklyn Bridge both on his 1866 Ohio River Bridge at Cincinnati (recently renamed the John A. Roebling Bridge) and on the ideas he articulated in his 1867 final report for that completed structure. In addition, Roebling wrote a brief design report for the Brooklyn Bridge in 1867.[10] From all of these documents, we can find that the greatest U.S. structural engineer of the nineteenth century thought about the bridge scientifically in terms of gravity loads and the effects of wind and traffic, socially in terms of the construction procedure and the uses of the bridge for mass transit, and finally symbolically as a great work of art and a place for people to promenade.

The suspension bridge as developed first by Telford makes possible the longest spans. The curving cable, being flexible, can only carry vertical loads by pure tension (the pulling force on a rope). We recall that the formula for the tension force at the midspan in a symmetrical cable (where the cable is the lowest) is $H = qL^2/8d$ where H is the tension in kips (1 kip = 1,000 lbs., or 1 kilopound), q is the vertical load in kips per ft. of bridge length, and *d* is the vertical distance from an imaginary line between the tower tops and the lowest point of the cable. The distance *L* is the span length between towers. For example, for the Brooklyn Bridge $q = 10$ kips per ft., $L = 1{,}595$ ft., and $d = 131$ ft., so that $H = 24{,}300$ kips.

We recall that this crucial formula, and various modifications of it, are the basis of nearly all major structural forms. In the suspension-bridge form it tells us that by increasing the cable sag (greater *d*) the cable tension will decrease, and hence less steel is needed. But one result is that

The Brooklyn Bridge

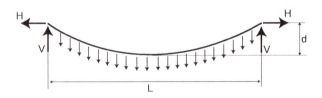

$$H = \frac{qL^2}{8d}$$

Horizontal component of the axial force in a cable

the towers now need to be higher. Thus the designer must weigh these two primary parts and decide on a form that is efficient (uses the least materials) and is also economical (uses the least money to build and maintain).

A second scientific factor is the wind. Roebling understood very well that many nineteenth-century bridges had been destroyed by wind storms that "were caused by the undulations of a very flexible floor, the rise and fall of which produced a succession of shocks leading to the necessity of a stiff floor, which of itself will prevent short undulations."[11] In a later report he amplified on the need for stiffness as opposed merely to bridge weight, which his chief rival, Charles Ellet, thought sufficient. But Ellet's Wheeling Bridge over the Ohio River collapsed in 1854, and as Roebling noted: "Weight is a most essential condition where stiffness is a great object, provided it is properly used in connection with other means. If relied upon alone as was the case in the plan of the Wheeling Bridge (U.S.), it may become the very means of its destruction. That bridge was destroyed by the momentum acquired by its own dead weight when swayed up and down by the force of the wind."[12]

Roebling then goes on to describe how a deck of great weight will not be much affected by moving traffic loads, but, if it has no other form of stiffness, will be in danger of oscillation from the wind. Thus, "although the weight of a floor is a very essential element of resistance to high winds it should not be left to itself to work its own destruction. Weight should be, simply, an attending element to a still more important condition viz: stiffness."[13] For Roebling, that stiffness was a combination of weight, a floor truss, and diagonal stays radiating out from the tower tops. These stays, were, however, ambiguous with respect to the gravity loads, and they tended to carry more such load than Roebling assumed. Hence they would break from time to time, which, to avoid endangering the bridge, required continual maintenance.

Roebling's Design: Social and Symbolic Perspectives

The primary social perspective in public works is cost, because that comes from public funds. Roebling had made long-span suspension bridges economical by developing in the early 1840s a method of spinning the cables in place. Wheels running over guide wires pulled the cable wire from one anchorage, over the tower saddles, to the other anchorage. Roebling's method was so sound that it is still used today and formed the basis for the 1980 Humber Bridge in Britain, now the world's longest-spanning bridge (1996).

Figure 11.2 John Augustus Roebling (1806–1869)

The very size of the bridge project and consequently the large sums of money involved made it a political object during construction. One of the bridge commissioners, Abram Hewitt, made a resolution in September 1876 to the effect that the Roebling Company be eliminated from supplying bridge cable because Washington Roebling, John Roebling's son, was chief engineer for the bridge (having replaced his father when the elder Roebling died in 1869). Thus the contract was awarded to J. Lloyd Haigh, who not only put defective wire in the bridge cables but also gave a substantial kickback to Abram Hewitt.[14]

John Roebling knew that his bridge would be a great monument, for as he wrote in 1867, "The great towers . . . will be ranked as national monuments, as a great work of art, and a successful specimen of advanced bridge engineering."[15] Once the bridge was opened on May 24, 1883, it was immediately recognized as the work of art Roebling claimed it to be.

But it was also criticized. Montgomery Schuyler, an architectural critic, wrote the first essay in the United States devoted to a work of pure engineering as a monument. Schuyler imagined a future archaeologist surveying our lost civilization from one of the towers of the Brooklyn Bridge:

> It is not unimaginable that our future archaeologist, looking from one of these towers upon the solitude of a mastless river and a dispeopled land, may have no other means of reconstructing our civilization than that which is furnished him by the tower on which he stands. What will his judgments of us be?
>
> It so happens that the work which is likely to be our most durable monument, and to convey some knowledge of us to the most remote posterity, is a work of bare utility; not a shrine, not a fortress, not a palace but a bridge. This is in itself characteristic of our time.[16]

Schuyler provided a sharp critique of the bridge, decrying its tower while praising its light suspension system:

> This aerial bow, as it hangs between the busy cities, "curving on a sky imbued with color," is perfect as an organism of nature. It is an organism of nature. There was no question in the mind of its designer of "good taste" or of appearance. He learned the law that struck its curves, the law that fixed the strength and the relation of its parts, and he applied the law. His work is beautiful.[17]

This is a clear opinion of the relationship between elegance and performance, but Schuyler goes on to condemn the towers by stating that the work "is almost unfailingly ugly when he does what he likes for the sake of beauty."[18]

Schuyler believed that John Roebling should have had an architect design his towers to reflect more of their load-carrying function. But this critic, essentially a literary man, did not realize that Roebling had designed those massive stone structures, by their weight, to sink the hollow caisson during construction. They were a part of the construction plan as well as serving, for Roebling, as gateways to the two civic centers linked by the bridge.

The major flaw in Schuyler's understanding lies in his erroneous belief that the proportions of the span, cable sag, tower height, deck dimensions, and stay locations are merely organisms of nature. This common fallacy associates engineering design with the automatic following of nature's laws. Roebling knew better and so do all the best engineering designers. The entire design of the Brooklyn Bridge is a unique product of Roebling's imagination. No other bridge, before or after, ever looked like this bridge. The bridge became and has remained a great symbol because of John Roebling's design, in the same way the Corliss engine came from the design ideas of one person.

George H. Corliss's Design: Scientific Aspect

George H. Corliss (1817–1888) had been one of Rhode Island's two commissioners to the Centennial Fair. In 1874 he had proposed to build an engine to power the fair but the executive committee, not liking to give the entire power contract to one person, solicited bids from numerous other engine suppliers. Unsatisfied with those results, they finally turned to Corliss in June 1875 and gave him the contract, which he fulfilled in less than a year.[19]

Figure 11.3 George Henry Corliss (1817–1888)

At the centennial Corliss's giant engine quickly became an icon, a word that in art history terms is often used to denote "an object of uncritical devotion, one which has an emotional meaning that transcends literal significance. Contemporaries clearly perceived the symbolic significance of the mammoth engine, which had been conceived [by Corliss] as a monument to—and incontrovertible proof of—America's technological maturity."[20]

None of the literary or artistic leaders who recognized the engine as an icon spoke of the actual operating ideas of Corliss's design. The engineering profession in 1876 certainly did so, and many in it realized that the Corliss engine did not represent the most advanced mechanical engineering. Everyone agreed, however, that the engine deserved the attention it got because of the powerful way it symbolized the United States. Yet some of those who, at first, sang its praises as a great aesthetic, even religious, object soon had other reactions. Indeed, the less one understood of the workings of the engine, it almost seems, the more one was susceptible to swings of elation and gloom over the prospect of a society dominated by such industrial giants. How did it really work? Again we have already explained this working, which we review here.

The simple equation that explains the heart of its working gives its power in the Hp formula of James Watt:

$$\text{Hp} = \frac{PLAN}{33,000}$$

In this formula, P, the pressure acting against the piston, was in the Corliss engine something like 45 lbs. per sq. in. due to the steam. The stroke or length of travel of the piston in the cylinder, L, was 10 ft. in the Corliss engine. The cross-sectional area of the piston on which the pressure acts, A, was 1,260 sq. in. for the 40-in. diameter cylinders of the engine, and the number of power strokes per min. made by the piston, N, was 72. Thus for each of the two cylinders in the Corliss engine, the power was

$$\text{Hp} = \frac{45 \times 10 \times 1,260 \times 72}{33,000} = 1,240 \text{ Hp}$$

We have already compared this formula to the one for bridges:

$$H = \frac{qL^2}{8d}$$

where we noticed several similarities and several crucial differences that help to clarify the complementary nature of structures and machines. First, the terms L and A define approximately the interior volume of the cylinder, which means they give the form of the heart in the engine. In the same way, L/d defines approximately the shape of the cable and thereby the proportion between cable span, cable sag, and tower height; that ratio gives the form of the central features of the structure.

Also, P in the machine and qL in the structure represent loads that act on the form to produce effects either of work in the machine or of force in the structure; but these effects are fundamentally different. In the machine the term N introduces time and motion, whereas in the structure no such term exists and the formula, therefore, expresses a timelessness that characterizes the relative permanence of structures.

On the other hand, N, measuring speed of operation, reflects relative permanence also: Where N is small there will be less wear on the parts, and indeed many old engines did not wear out but became uneconomical and hence obsolete.

The most striking difference in the two formulas, however, is that between the results: H, a tension force (lbs.) in the cable at midspan (and nearly a constant throughout the full cable length), and Hp, a measure of power delivered by the piston to the flywheel and hence to the entire fair; the static cable suspended in tension from anchorage to anchorage versus the dynamic shafting delivering power throughout a vast complex of belts. The extent of the bridge is immediately perceived visually as a whole, whereas the extent of power is felt, is imagined, but cannot be visually taken in with one view. The cable is a line of force and its repose gives it its iconic value whereas the cylinder piston is a motion of force and its reciprocations impress on the viewer the promise of unseen and far-reaching power. Of course, with electricity this view of the machine, as dynamo, would become even more widely felt and more dramatically expressed by reflective people.

The Corliss Engine: Its Social and Symbolic Perspectives

The machine was transforming U.S. society with such dazzling speed that the nation rose from the world's fourth-ranked manufacturer in 1860 to the world's leader three decades later. By 1894 the value of manufactured products of the United States was nearly as great as those of Great Britain, France, and Germany put together.[21]

From 30,000 miles of track before the Civil War, U.S. railroads rose to 166,000 miles by 1890. The Corliss engine stood not just for stationary power plants but also for the wildly accelerating mobility all pushed by steam. The nation symbolized by the centennial was "welded into a unified network. The nation had developed a truly national economy."[22] Workers were becoming increasingly specialized and more productive; if they lost their independence because of the machine, they gained more job opportunities in the rapidly expanding economy. Business became more competitive but also more complex, all growing from the inherent nature of machine industry.

Technology transformed the city. "Graceful new skyscrapers and majestic bridges, marvels of art and engineering, towered beside crumbling slum tenements." The promise of more beautiful cities was compromised by the presence of poverty. Still, the city was a powerful magnet, a dynamo, that "attracted the most ambitious, intelligent, and imaginative persons the nation produced."[23]

Government also became more complex, partly because of industry and partly because of new public works. The trend of government regulation that began in the steamboat era would continue with efforts to regulate the new giant industries that emerged in the late nineteenth century.

Woodrow Wilson, in one of the first Ph.D. dissertations written in the United States on political science, criticized the House of Representatives as a weak, bumbling, decentralized instrument and recommended a more efficient and centralized operation. "The more power is divided, the more irresponsible it becomes," he wrote in 1885, and that image fit directly with the image of the Corliss engine in Philadelphia.[24] Wilson could not escape the implications of a responsible power that technology was making clear. Politics, some would say, was catching up with technology.

George Corliss did not design his engines merely to be an efficient powerhouse; rather he had a deep belief in "honest work" and he had a "flair for showy mechanisms and for monumentality." This belief meant that his engines were extraordinarily reliable and durable. It was said that "no Corliss engine was ever worn out." Instead, "due to changed economic conditions rather than mechanical failure," the engines were simply retired.[25]

Early in the nineteenth century, machinery made in the United States had the reputation for being loosely made and short-lived. Corliss changed all that in a most unusual way. His basic patent of 1849 was ostensibly directed to a new type of steam valve on the cylinder, but more properly his idea was to build a firm and elegant machine. Its looks were to be its hallmark, and this meant to Corliss an engine that was both more efficient and more quiet. The unusual feature of the Corliss engine was its four-valve system for controlling the steam. This complex mechanism was visible, operating on the outside of the cylinder through an oscillating wrist plate. This plate, through an intricate and delicate series of links and wheels, operated each of the four steam valves one after the other and at just the right moment. Here was the aesthetic of the mechanical that people stood watching, fascinated. But this four-step steam play cycled almost without any sound at all. It was mime and ballet albeit entirely predictable. The great engine was under the control of Corliss's little links. It was both a finely tuned instrument and a grandly massive monument. William Dean Howells, a leading literary figure and editor of the *Atlantic Monthly*, caught its spirit in writing:

> The Corliss engine does not lend itself to description; its personal acquaintance must be sought by those who would understand its vast and almost silent grandeur. It rises loftily in the center of the huge structure [Machinery Hall], an athlete of steel and iron with not a superfluous ounce of metal on it; the mighty walking-beams plunge their pistons downward, the enormous fly-wheel revolves with a hoarded power that makes all tremble, the hundred life-like details do their office with unerring intelligence. In the midst of this ineffable strong mechanism is a chair where the engineer sits reading his newspaper, as in a peaceful bower. Now and then he lays down his paper and clambers up one of the stairways that cover the frame-work, and touches some irritated spot on the giant's body with a drop of oil, and goes down again and takes up his newspaper.[26]

This direct observation of the engine and the engineer was not the only type of response to the machine as monument. The actual working of the machine "was not the source of its overwhelming popularity and acclaim in the nineteenth century. Instead, visitors to the Centennial were chiefly impressed by the engine as a powerful, indeed monumental, symbol of man's technological triumphs and a titanic form to inspire the romantic imagination. Detailed knowledge of mechanical functions was not only superfluous for these seekers of the technological sublime, but actually threatened to impair the free flow of associations."[27]

The impression gotten from this quotation is that the valves and cylinders merely set off romantic thoughts and that the engine itself was slow and not efficient. But this view, appealing to that writer, is far from the truth. It is true that the steam turbine was already making the reciprocating engine obsolete in principle for central power plants after 1876. But it was demonstrably untrue that the Corliss engine itself was thus more a monument than an object of utility. Corliss engines were used throughout the world in great numbers up to the first years of the twentieth century.

Moreover, the New York City system of rapid transit (elevated surface and subway lines) took power from such engines, with "the ultimate installation [being] in subway trains of the Interborough Rapid Transit. Nine pairs of engines, each as tall as the Centennial engine, were installed in a cathedral-like engine hall. Between each pair of engines was a directly-connected 5000-kilowatt generator whose rotor acted as a flywheel for the engines. The engines ran for more than fifty years, being scrapped finally around 1960."[28]

In short, Corliss engines were competitive throughout the nineteenth century,[29] they were visual monuments because of their designer's unique aesthetic sensitivity, and they were not decorated by extraneous "artistic" details. Corliss, as engine designer and builder, ranks with Roebling as an engineering artist who made, out of the means of efficiency and economy, objects of engineering that became great cultural symbols of their era thanks to their designer's insistence on visual elegance.

Corliss's Last Stand

The fair stands outside of main political events, but it characterized the mechanization of industry and, with the Corliss engine, an approach to engineering that bordered on reverence. Indeed the period between 1876 and 1893, the year of the Chicago Columbian exposition (one year late in celebrating the 400th year of Columbus's voyage), began the rise of the great U.S. industries whose power set the Bostonian Henry Adams thinking about the religious nature of machinery.[30]

However impressed visitors were by his engine, George Corliss could not find a buyer for it at the end of the centennial in 1876, so he took it apart and shipped it back to Rhode Island. After four years of repose the engine found a home when George Pullman (1831–1897), the pioneer of railroad sleeping cars, bought it to power his model town of Pullman, Illinois, then under construction. The town was designed by the architect Solon S. Berman, who wrote years later:

> From its first inception in the mind of George M. Pullman it was his idea to make the town a model of its kind and the works not only productive but a showcase in which all might take pride. This led Mr. Pullman to purchase the great Corliss engine which furnished the power for the Centennial Exposition in Philadelphia in 1876 and place it in a conspicuous building, the so-called Machinery Hall, through the large windows of which the machine could be seen in operation by passersby and by passengers on the trains of the Illinois Central and Michigan Central Lines.[31]

It took 35 boxcars to ship the great engine to Pullman and on April 2, 1881, it was turned on once again by dignitaries. George Pullman knew he had bought an icon.[32]

Just as at Machinery Hall in 1876, George Pullman put the Corliss engine in a prominent place at Pullman in a building facing an artificial lake, Lake Vista, fed by condensing steam from the engine itself. The engine was mounted on a platform and set in the context of elegant hardwood. Pullman had bought the Corliss engine in 1880 to provide the power for his car factory. He had wanted this immense machine also to be a landmark for his model town. Even though the engine was, by 1880, not the state of the art, Pullman's decision was economically sound— he "got about thirty years of relatively trouble-free service from the Corliss, whose task was to propel more than 3 miles of lineshafting."[33] Pullman characterized the end of an era: mechanical transmission of power and the paternal governance of a factory town. But his influence on the railroads lasted long into the twentieth century and far into nations around the world.

In 1883 the railroads forced the nation into the last of all fundamental connections by organizing the country into four time zones. Building upon the gradual standardization of track width, the railroads left a permanent legacy of order by which we now regulate national life. At the same time, collectively speaking, the companies themselves remained disorganized and in sharp competition. There was no national system but rather a series of independent lines. This would be one characteristic of the U.S. democratic experiment: a continual ambiguity between central planning and local initiative, between federal control and private enterprise.

As the older dominance, symbolized by the Brooklyn Bridge and the Corliss engine, receded and the newer ones grew impressively, the industrial world began to recognize a pattern of change that has persisted toward the end of the twentieth century. It appears to be related to the modern engineering culture.

From the late-twentieth-century perspective, we can see the nineteenth-century changes as the first half of a 200-year transformation of society within which some patterns emerge. These patterns reveal how society itself has evolved from rural and agrarian to urban and industrial.

From American independence to the early nineteenth century, there was little change although much isolated experimentation with steam engines and small textile mills. The first major transformations began in 1807 with Fulton's success on the Hudson River and culminated in the 1820s with the opening of the Erie Canal and the financial prowess of Lowell. Steamboats, canals, and textiles remained dominant up to midcentury, but after 1830 the railroad and its allied iron industry began more and more to shape the nation. By the end of the Civil War the steamboat, the canal, and the New England textile industry were in decline as the railroad dominated.

Starting after the Civil War, and in rapid succession, iron gave way to steel, steam power began to generate electricity, and the oil industry arose. When the Brooklyn Bridge opened on May 24, 1883, the great U.S. industries were beginning their ascendancy. Yet the older engineering works did not disappear. The greatest canal was yet to be built (Panama), the textile industry kept growing, and steam-powered ships would reach titanic sizes. What began with the industrial revolution has remained, however modified, and seems likely to do so for a long time. New transportation vehicles would come, new industries would begin, and new structures would result from the new materials of structural steel and reinforced concrete.

Visitors to the Centennial Exhibition marveled at the machinery and the products of the United States. Perceptive foreign delegations recognized to their amazement that the young nation was the world leader in engineering. The explanations and the debates had begun over why this was so and over what it meant for society.

Andrew Carnegie had his own explanation. In 1886 he published *Triumphant Democracy*, in which he argued that general prosperity and general culture had both reached such a high level that the United States

> is no longer compelled to rest its claims for recognition upon its vast material resources. It now challenges the older nations in the higher domain of intellectual, scientific, and artistic development.[34]

This had been Carnegie's ideal as he grew more and more wealthy. His wealth, he believed, was "a trust to be administered during life for the good of the public from whom it came."[35] Ever since the 1880s, observers have debated Carnegie's optimistic view of capitalism, free enterprise, and democracy in the United States. Because the wealth throughout the nation was so great and spread to such a large percentage of the population, the poverty that persisted aroused critics and reformers. They disputed Carnegie and some proposed socialism, with a few even introducing Marxism.[36] The most important fact about this debate was its very existence. In a free democratic society, such divergent views kept alive the tension between private initiative and public welfare that had informed the congressional debates over steamboat regulation. Without regulation there was a danger to the traveling public; without the steamboat and the subsequent engineering innovations, there could have been no traveling public.

By 1883 society was already transformed; in another century that transformation would continue, built upon the broad and sturdy foundations laid by steamboats, turbines, textiles, telegraph, railroad, steel, and electric power: the works of engineering that made America modern.

Notes and References

1. Modern Engineering and the Transformation of America

1. Brooke Hindle, *Emulation and Invention*, New York, 1981, p. 97.
2. Earl A. Powell, *Thomas Cole*, New York, 1990, pp. 26, 36, 113, and 118.
3. Othmar H. Ammann, "George Washington Bridge: General Conception and Development of Design," *Transactions*, American Society of Civil Engineers, Vol. 97, 1933, p. 39.
4. H. W. Dickinson, *A Short History of the Steam Engine*, Cambridge, England, 1939.
5. This is a theoretical statement, as are all such simplified explanations of the formulas in this text. The characteristics of a cylinder, the way the steam is admitted, and the means by which power is transmitted from the cylinder all influence the useful power produced.
6. D'Arcy Thompson, *On Growth and Form*, Cambridge, 1966 (originally published in 1917).

2. Watt, Telford, and the British Beginnings

1. H. W. Dickinson, *James Watt: Craftsman and Engineer*, Cambridge University Press, 1936, p. 200.
2. David P. Billington, *The Tower and the Bridge: The New Art of Structural Engineering*, Princeton University Press, 1985 (originally published by Basic Books, New York, 1983), pp. 27–44.
3. Thomas Telford, "Bridge," *New Edinburgh Encyclopedia*, 2nd American edition, Vol. 4, Whiting and Watson, New York, 1814, pp. 470–532.
4. T. S. Ashton, *Iron and Steel in the Industrial Revolution*, Manchester, 3rd edition, 1963, p. 35.
5. *Encyclopedia Britannica*, Vol. 12, 1963, p. 308.
6. Ashton, op. cit., p. 64.
7. H. W. Dickinson, *A Short History of the Steam Engine*, Cambridge University Press, 1939, pp. 68–69. See also Dickinson, op. cit., 1936, p. 36.
8. Dickinson, op. cit., 1939, p. 62. The calculations that follow are those found in John Farey, *A Treatise on the Steam Engine,* London, 1827, pp. 172–173.
9. Dickinson, op. cit., 1939, plate X. The work of a pumping engine was called "duty," which was measured in pounds of water raised 1 ft. by the use of 1 bushel of coal (84

lbs.). Newcomen's 1718 engine gave a duty of 4.3 million. John Smeaton (1724–1792) improved this by 1774 to 12.5 million, and Watt's engines gave 22.6 million by 1779 and 39 million by 1792. Further improvements in the nineteenth century led to a duty of 203 million by 1906.

10. F. M. Scherer, "Invention and Innovation in the Watt-Boulton Steam-Engine Venture," *Technology and Culture: An Anthology*, Eds. Melvin Kranzberg and William H. Davenport, New American Library, 1972, p. 305. This article was published originally in *Technology and Culture*, Vol. 6, No. 2, 1965, pp. 165–187.

11. By 1800 the firm had built roughly 500 engines of which 62 percent were rotative, mostly used in the textile industry. Boulton was clearly right in his judgment of the need for rotative power. See Dickinson, op. cit., 1939, p. 88. In 1781 Boulton had written that "the people in London, Manchester, and Birmingham are *steam mill mad*." Ibid., p. 80.

12. One of Watt's defects as a practical designer was his continual urge to try a multitude of different devices or approaches to the many parts of his engines.

13. Eric Robinson and A. E. Musson, *James Watt and the Steam Revolution*, Adams and Dart, London, 1969, pp. 1–21.

14. Thomas Telford, *Life of Thomas Telford, Civil Engineer*, London, James and Luke G. Hansard and Sons, 1838, p. 34. Edited by John Rickman and published after Telford's death.

15. J. B. Lawson, "Thomas Telford in Shewsbury: The Metamorphosis of an Architect into Civil Engineer," *Thomas Telford: Engineer*, Ed. Alistair Penford, Thomas Telford Ltd., London, 1980, p. 13.

16. The primary force in an arch bridge is axial and is approximately proportional to the span length squared. To keep the same safety, the arch material should be increased by the same amount. Thus, if the 100-ft. span Iron Bridge used 30 tons of iron, then a span of 130 ft. should have used $(30) \times (130)^2/(100)^2 = 50.7$ tons, whereas Telford used only $0.5 \times 30 = 15$ tons. Thus by comparison Telford's efficiency is $50.7/15 = 3.4$ times that of the Iron Bridge.

17. The London Bridge design is discussed in detail in A. W. Skempton, "Telford and the Design of a New London Bridge," *Thomas Telford, Engineer*, pp. 62–83.

18. Ibid., pp. 79–80.

19. Roland A. Paxton, "Menai Bridge 1818–26: Evolution of Design," *Thomas Telford, Engineer*, pp. 84–86.

20. John R. Hume, "Telford's Highland Bridges," *Thomas Telford, Engineer*, p. 164.

21. For a detailed discussion of Telford's aesthetic ideas, see Billington, op. cit., Chapter 2.

22. Clarence V. Knudsen, "The New River Gorge Bridge: World's Longest Steel Arch,"*Civil Engineering*, Vol. 47, No. 2, February 1977.

23. Dickinson, op. cit., 1939, Chapter 10. "The Philosophy of the Steam Engine," pp. 173–182.

24. Billington, op. cit., pp. 43–44.

25. C. L. M. H. Navier, *Memoire sur les Ponts suspendus*, Paris, 1823.

26. Quoted in Dickinson, op. cit., 1936, pp. 53–54.

27. Joseph Hariss, *The Tallest Tower: Eiffel and the Belle Epoque*, Boston, Houghton Mifflin, 1975, p. 10.

28. These ideas are discussed by R. G. Collingwood, *Essays in the Philosophy of History*, Ed. William Debbins, University of Texas Press, 1965. See especially the essays "The Theory of Historical Cycles" and "A Philosophy of Progress."

29. Alexander Gibb, *The Story of Telford,* Alexander Mechose, London, 1935.
30. Not all Watt engines had been scrapped by 1900, but none were still serving their original purposes in competitive industrial contexts. However, unlike much modern machinery, the Watt engines did not usually wear out; because they operated at such slow speeds and low pressures, they simply became uneconomical to run. Economy, of course, had been their justification in the first place.
31. Telford, op. cit., 1814, pp. 479–489.

3. Fulton's Steamboat and the Mississippi

1. Louis C. Hunter, "The Heroic Theory of Invention," *Technology and Social Change in America*, Ed. Edwin T. Layton, Jr., New York, 1973, p. 25.
2. Eugene S. Ferguson, "Steam Transportation," *Technology in Western Civilization*, Vol. 1, Eds. M. Kranzberg and C. W. Pursell, Jr., 1967, pp. 286–291. For a recent article that stresses the importance of Fitch, see C. M. Harris, "The Improbable Success of John Fitch," *Invention and Technology*, Vol. 4, No. 3., Winter 1989, pp. 24–31.
3. John S. Morgan, *Robert Fulton*, New York, 1977, pp. 106–107.
4. Ibid., p. 111.
5. Robert Fulton, *Treatise on the Improvement of Canal Navigation*, London, 1796.
6. Quoted in Morgan, Ibid., p. 132.
7. Brooke Hindle, *Emulation and Invention*, New York, 1981, p. 52.
8. This trip is well described by Cynthia Owen Philip, *Robert Fulton: A Biography*, Franklin Watts, New York, 1985, pp. 200–204.
9. Hindle, op. cit., p. 56.
10. Ibid., p. 135.
11. Robert Fulton, *"Patent Application,"* January 1, 1809, p. 2.
12. Ibid., p. 3 and pp. 10–14. There is an obvious error in Fulton's calculation for 6 mph. With that error corrected, we find his results are reasonably close to a variation following V^2.
13. In an 1759 paper, John Smeaton had described his tests showing that undershot waterwheels (essentially what Fulton's paddlewheels were) had only half the power of overshot ones. In other words, the efficiency of the paddles would only be 50 percent. Hence for Fulton, whose paddles acted as undershot wheels in reverse, the power provided had to be double the power required; i.e., the power required was TV_T, where $V_T = 4$ mph and the power supplied was TV_P, where V_P the paddle velocity was taken by Fulton to be $2V_T$. See Hunter Rouse and Simon Ince, *History of Hydraulics*, Dover, pp. 121–122.
14. Mark Beaufoy, *Nautical and Hydraulic Experiments*, London, May 1834. This 688-page document contains detailed test results. Fulton took data on drag from Tables I and III as the basis for his design. He had already used these data in 1803. See Cynthia Owen Philip, *Robert Fulton: A Biography*, New York, 1985, pp. 142–143.
15. Fulton, op. cit., 1809, p. 3. The Beaufoy tests were not reliable for ship design. The late twentieth-century way to determine the drag coefficient is through the use of two nondimensional ratios: Reynold's number and Froude's number. For a boat of Fulton's design moving at such slow speeds, Froude's number is so small that it would not enter here; but Reynold's number

$$R_N = \frac{L \times V}{v} = \frac{154 \text{ ft.} \times 5.87 \text{ }^{ft}\!/_{sec}}{1.22 \times 10^{-5} \text{ }^{ft^2}\!/_{sec}} = 7.41 \times 10^7$$

where ν is the kinematic viscosity which for water at 60° F is 1.22×10^{-5} ft^2/sec. For this value of R_N, the skin friction (on the wetted surface of the boat) would have a drag coefficient C_D of about 0.0022 (see *Principles of Naval Architecture, Second Revision*, Vol. II, Ed. E. V. Lewis, Jersey City, N.J., p. 12).

For the paddle wheels, Fulton's results lead to a drag higher than would be used today where square plates have a coefficient of about 1.1; see R. C. Binder, *Fluid Mechanics*, New York, 1943, p. 137. Also there is not always a paddle surface perpendicular to the direction of motion so that the average thrust is less than Fulton assumed.

16. Vannevar Bush, *Science, the Endless Frontier: A Report to the President*, Washington, DC, 1945, pp. 13–14, cited in Edwin T. Layton, Jr., "American Ideologies of Science and Engineering," *Technology and Culture*, Vol. 17, 1976, p. 689.

17. Philip, op. cit., pp. 142–153.

18. Hunter, op. cit., p. 25–26.

19. Philip, op. cit., p. 120.

20. T. S. Ashton, *Iron and Steel in the Industrial Revolution*, Manchester, 3rd edition, 1963, p. 35.

21. G. Bathe and D. Bathe, *Oliver Evans*, Philadelphia, 1935, p. 115.

22. Robert Hare, "Letter to the editor," *The Franklin Journal and American Mechanic's Magazine*, Vol. 2, 1826, p. 147. See also G. Bathe and D. Bathe, *Oliver Evans*, Philadelphia, 1935, pp. 232–238.

23. E. Gordon, *Structures, or Why Things Don't Fall Down*, New York, 1978, pp. 117–123.

24. The material itself could carry 6,150 psi without cracking. Evans had determined experimentally that wrought iron could carry 64,000 psi and thus his design should have been very safe. However, the heat applied to the boiler walls and the rivet holes was often enough to cause failure where tension stresses were the highest.

25. Hunter, op. cit., p. 34.

26. Ibid., p. 35. See also Edith McCall, *Conquering the Rivers: Henry Miller Shreve and the Navigation of America's Inland Waterways,* Baton Rouge, 1984.

27. John G. Burke, "Bursting Boilers and Federal Power," *Technology and Culture*, Vol. 7, No. 1, Winter 1966, p. 18. See also Layton, op. cit., 1973, p. 115. For the *Washington* explosion, see McCall, op. cit., pp. 145–147.

28. Ibid., p. 100.

29. Ibid., p. 116. See also "Alfred Guthrie," *Dictionary of American Biography*, Vol. 8, pp. 59–60.

30. L. Stebbins, pub., *Eighty Years' Progress of the United States*, New York, 1864, p. 243. Quoted in Burke, op. cit., p. 118.

31. Burke, Ibid., pp. 118–119.

32. Ibid., p. 119.

33. Mark Twain, *Life on the Mississippi*, New American Library, 1980, pp. 135–136 (first published in 1883).

4. Lowell and the American Industrial Revolution

1. Ferris Greenslet, *The Lowells and Their Seven Worlds*, Boston, 1946, p. 160.

2. Ibid., p. 76. The old judge's great-great-great-great grandson, John Lowell (b. 1919), still played an active role in Harvard and other educational institutions around Boston in the late 1980s. See *Who's Who in America*, 44th ed., 1986–87, Vol. 2, p. 1737.

3. The account of F. C. Lowell's time abroad appears in Greenslet, op. cit., pp. 125–130, and in Robert F. Dalzell, Jr., *Enterprising Elite: The Boston Associates and the World They Made*, Cambridge, Mass., 1987, pp. 5–25.

4. Ibid., pp. 12–13.

5. Quoted in Dalzell, Ibid., p. 12.

6. Howard P. Segal, *Technological Utopianism in American Culture*, Chicago, 1985, pp. 63–67.

7. Louis C. Hunter, *Water Power*, Charlottesville, 1979, pp. 459–460.

8. Greenslet, op. cit., p. 157.

9. Dalzell, op. cit., pp. 26–30.

10. Greenslet, op. cit., pp. 156–160, and Dalzell, op. cit., pp. 30–36.

11. Harry C. Dinmore, "Proprietors of Locks and Canals: The Founding of Lowell,"*Cotton was King: A History of Lowell, Massachusetts*, Ed. Arthur L. Eno, Jr., Lowell, 1976, p. 69.

12. James B. Francis, *Lowell Hydraulic Experiments*, Boston, 1855, Introduction, p. x.

13. Dalzell, op. cit., pp. 84 and 88. See also John A. Goodwin, "Villages at Waresit Neck," in Eno, op. cit., pp. 59–61.

14. Dalzell, op. cit., p. 48.

15. Patrick M. Malone and Larry D. Lankton, *Water Power in Lowell, Massachusetts*, presented to the Subcommittee on National Parks and Recreation to the House Committee on Interim and Insular Affairs, August 19, 1974, pp. 1–11.

16. Patrick M. Malone, *Canals and Industry: Engineering in Lowell, 1871–1880*, Lowell Museum, Lowell, Mass., 1983, p. 5.

17. Ibid., p. 20. See also F. B. C. Bradlee, *The Boston and Lowell Railroad etc.,* Salem, 1918, p. 9.

5. Francis and the Industrial Power Network

1. "Memoirs of James Bicheno Francis," *Proceedings of the American Society of Civil Engineers*, Vol. 18, 1872, pp. 74–88.

2. Terry S. Reynolds, "The Emergence of the Breast Wheel and Its Adoption in the United States," *The World of the Industrial Revolution*, Lowell Conference on Industrial History, 1984, Ed. Robert Wieble, 1986, pp. 76–77.

3. Patrick M. Malone and Larry D. Lankton, *Water Power in Lowell, Massachusetts*, presented to the Subcommittee on National Parks and Recreation to the House Committee on Interim and Insular Affairs, August 19, 1974, pp. 1–11.

4. James B. Francis, *Lowell Hydraulic Experiments*, 3rd ed., New York, 1871, pp. 148–154. The first edition was in 1855. The second edition was 1868.

5. Malone and Lankton, op. cit., p. 19.

6. Ibid., p. 29.

7. Malone, 1983, op. cit., p. 5, for the Pawtucket Canal, and for the Northern Canal, p. 15. The Hp in 1853 comes from Dinmore, op. cit., p. 77, where the power is given as 139 11/50 mill powers. Because one mill power is 85.2 Hp (see Hunter, op. cit., p. 211), the total Hp = 139 11/30 × 85.2 = 11,900 Hp.

8. It ran about 4,400 ft. with a drop of 3 ft. for a slope of $S = 3/4{,}400 = 0.00068$. The canal has stone walls and bottom, permitting a relatively smooth water flow. The standard formula for open channel flow, called Manning's formula, states that flow increases with the slope S (as the square root, or $\sqrt{S}$), with the area A, and with the ratio r of the area to the wetted perimeter (as the cube root of r squared or $r^{2/3}$) and that

flow decreases with the friction n of the moving water against the wetted perimeter of the canal. Thus for the Northern Canal flowing less than half full, $D = 7$ ft., the area $A = 700$ sq. ft., $r = 700/(14 + 100) = 6.15$ ft., $S = 0.00068$, and for stone $n = 0.025$ so that

$$Q = \frac{1.486}{n} r^{2/3} S^{1/2} A = 59.44 \times 3.35 \times 0.026 \times 700 = 3,620 \text{ cu. ft./sec.}$$

The term 1.486 is a constant required for the ft.-lb. units. The Manning equation as given with that constant is only valid in the ft.-lb. system. See, for example, *Civil Engineering Handbook*, Ed. L. C. Urquhart, 3rd ed., 1950, p. 368.

9. Malone, op. cit., p. 15.
10. Malone and Lankton, op. cit., p. 23.
11. Francis, *Experiments*, op. cit., pp. 2–3.
12. Ibid., pp. 6–7.
13. Ibid., pp. 32–33. Test number 1 occurred on February 17, 1851.
14. Ibid., p. 26.
15. Ibid., p. 38.
16. Ibid., p. 33. For test #30,

$$V = (0.851)(68) = 58 \frac{\text{ft.}}{\text{sec.}} \text{ or } \frac{58}{121.5} = .48 V_{\text{max}}$$

For this velocity the friction force $T = 1,525$ lbs. and hence the mechanical power

$$P_{\text{out}} = (1,525)(58) = 88,300 \frac{\text{ft.–lbs.}}{\text{sec.}}$$

The value of P_{in} measured was

$$111,220 \frac{\text{ft.–lbs.}}{\text{sec.}}$$

or slightly less than that in test #1. Francis actually plotted efficiency as the ordinate and a ratio of velocities as the abscissa. This ratio is the velocity of the interior circumference of the wheel divided by the velocity due to the head acting on the wheel. For test #1, for example,

$$H = 12.864 \text{ ft., giving } V = \sqrt{2gH} = 28.78 \text{ ft./sec.}$$

and with the interior circumference of 21.21 ft. ($R = 3.375$ ft.) and 0.894 revolutions per sec., the wheel velocity is $(21.21)(0.894) = 18.96$ ft./sec. Thus the ratio is

$$\frac{18.96}{28.78} = 0.66$$

As the head was nearly constant for all 92 tests, Francis's plot is essentially efficiency versus wheel velocity. For tests 1 through 50, H varied between 12.471 and 12.977.

17. Ibid., p. 52.
18. For a balanced discussion of the contributions of James Francis, see Louis C. Hunter, *A History of Industrial Power in the United States, 1780–1930: Volume One: Waterpower*, Charlottesville, 1979, pp. 328–342. A more detailed study is Edwin T. Layton, Jr., *From Rule of Thumb to Scientific Engineering: James B. Francis and the Invention of the Francis Turbine*, Stony Brook, N.Y., 1992. For the role of Francis, see also Layton, "Scientific Technology, 1845–1900: The Hydraulic Turbine and the Origins

of American Industrial Research," *Technology and Culture,* Vol. 20, No. 1 (Jan. 1970): pp. 64–89.

19. See, for example, R. C. Binder, *Fluid Mechanics*, Prentice-Hall, New York, 1943, pp. 102–103. Francis did not give the derivation probably because it requires the calculus, a subject he avoided in *The Experiments*. The derivation is from Bernoulli's theory: the ideal velocity for a fluid particle a distance z below the fluid surface, $V_z = \sqrt{2gz}$. Then the ideal flow would be $dQ = LV_z dz$, where dz is the infinitesimal thickness of the fluid at level z. Thus the total ideal flow

$$Q = \int_O^H dQ = \int_O^H LV_z \, dz$$

or

$$Q = \int_O^H L\sqrt{2g}z \, dz = \frac{2}{3} L\sqrt{2g} \, H^{3/2}$$

which we can interpret as the area HL times a new velocity

$$V = \frac{2}{3}\sqrt{2gH}$$

From this we find

$$C = \frac{2}{3} \sqrt{2g} = 5.35$$

theoretically, but the value Francis found from tests was 3.33.

20. Francis, op. cit., p. 74.
21. Ibid., p. 75.
22. Ibid., p. 86.
23. Ibid., p. 118.
24. Ibid., p. 118. The use of $b = 0.1$ appears in column 15 of table XIII on p. 123.
25. Ibid., pp. 122–125. Francis discusses the final choice of $C = 3.33$ on p. 119.
26. Ibid., pp. 126–129.
27. Ibid. pp. 133–135.
28. In a standard 1985 text on fluid mechanics the formula for weirs, for example, appears as (for $n = 2$):

$$Q = 0.59 \sqrt{g}(L - 0.2H)H^{3/2}$$

which gives for $C = 0.59\sqrt{g} = 3.35$, or almost exactly Francis's result of 1852. See Robert W. Fox and Alan T. McDonald, *Introduction to Fluid Mechanics*, John Wiley and Sons, New York, 3rd ed., 1985, pp. 546–550. Actually the formula for

$$C = 0.59 + (0.08) \left(\frac{H}{Zw} \right)$$

where Z_w is the height of the weir above the channel base. In a sample problem on p. 549 the authors assume $H \ll Z_w$ and hence $C = 0.59$. This assumption applies well to Francis's tests (see plate XIII of *The Experiments*). In 1950, a standard civil engineering handbook noted with respect to Francis's equation for including the effects of end contractions, $L - (0.1)nH$, that "no better general formula has, however, been suggested." See Horace W. King and Ernest F. Brater, "Hydraulics," *Civil Engineering Handbook*, Ed. L. C. Urquhart, 3rd ed., 1950, p. 338.

29. James B. Francis, "Experiments on the Deflection of Continuous Beams, Supported at Equidistant Points," *Transactions ASCE*, Vol. 1, 1872, pp. 117–118.

30. See, for example, Stephen P. Timoshenko, *History of Strength of Materials*, 1952 (reprinted 1983, Dover), McGraw-Hill, New York, p. 100.

6. The Stephensons, Thomson, and the Eastern Railroads

1. Louis C. Hunter, *A History of Industrial Power in the United States, 1780–1930, Volume One: Water Power in the Century of the Steam Engine*, Charlottesville, Va., 1979, p. 343.

2. Eugene S. Ferguson, "Steam Transportation," *Technology in Western Civilization*, Vol. 1, eds. M. Kranzberg and C. W. Pursell, Jr., Oxford University Press, 1967, pp. 294–297.

3. Oliver Jensen, *Railroads in America*, New York, 1975, p. 17.

4. J. G. H. Warren, *A Century of Locomotive Building by Robert Stephenson & Co.: 1823–1923*, Newcastle-upon-Tyne, 1923, p. 251.

5. E. T. MacDermot, *History of the Great Western Railway: Volume One, 1833–1863*, revised by C. T. Clinker, London, 1964, pp. 16–17.

6. W. P. Marshall, "Description of the Patent Locomotive Steam Engine made by Robert Stephenson and Co. of Newcastle-Upon-Tyne," *Locomotive Engines*, First Volume of the New Edition of *Thomas Tredgold on the Steam Engine,* London, 1850, Sixth Paper, 68 pages with four plates.

7. Ibid., pp. 1–2.

8. Ibid., p. 68.

9. Ibid., p. 11.

10. Ibid., pp. 13–14.

11. Ibid., p. 67. The engine appears to be the same one described as number 123 by Michael R. Bailey, "Robert Stephenson and Company: 1823–1836," master of arts thesis, University of Newcastle-upon-Tyne, June 1984, Appendix IV, p. 6.

12. John H. White, Jr., *A History of the American Locomotive, Its Development: 1830–1880*, New York, 1979 (first published in Baltimore, 1968), pp. 74–76.

13. Warren, op. cit., p. 85.

14. David P. Billington, *The Tower and the Bridge*, New York, 1983, pp. 49–50.

15. Alfred Pugsley, "Clifton Suspension Bridge," *The Works of Isambard Kingdom Brunel*, London, 1976, pp. 51–68.

16. L. T. C. Rolt, *Isambard Kingdom Brunel*, London, 1957, pp. 407–408.

17. MacDermot, op. cit., pp. 17–18.

18. Ibid., p. 33.

19. Ibid., p. 29.

20. Rolt, op. cit., p. 198.

21. Ibid., pp. 199–202.

22. Billington, op. cit., 1983, pp. 54–59.

23. Brooke Hindle and Steven Lubar, *Engines of Change: The American Industrial Revolution 1790–1860*, Smithsonian Institution Press, Washington, DC, 1986, Chapter 8, pp. 125–151.

24. Eugene S. Ferguson, "Steam Transportation," *Technology in Western Civilization*, Vol. 1, eds. M. Kranzberg and C. W. Pursell, Jr., Oxford University Press, 1967, pp. 298–301.

25. Hindle and Lubar, op. cit., pp. 145–148.
26. John A. Kouwenhoven, *Made in America*, Doubleday Anchor Book, 1962 (first edition 1948), pp. 30–33.
27. White, op. cit., pp. 269–279. Baldwin stated that his power was "equal to 30 Horses at 15 mph." Probably P actual was well below 120 psi and his power referred to the pulling of cars.
28. Carl Condit, *The Port of New York: A History of the Rail and Terminal System from the Beginnings to Pennsylvania Station*, Chicago, 1980, pp. 44–48 and pp. 50–51.
29. James A. Ward, *J. Edgar Thomson: Master of the Pennsylvania*, Greenwood Press, Westport, Conn., 1980, pp. 136–137.
30. Ibid., p. 221. See Max Ways, "A Hall of Fame for Business Leadership," *Fortune*, Jan. 1975, p. 71.
31. Ibid., p. 222.
32. Ibid., p. 44.
33. Ibid., pp. 70–71.
34. Ibid., p. 81.
35. Ibid., p. 42 and p. 81.
36. Ibid., p. 99, and Jensen, op. cit., pp. 36–39.
37. Ward, op. cit., p. 96.
38. Ibid., p. 107.
39. Ibid., p. 140.

7. Henry, Morse, and the Telegraph

1. John F. Stover, *The Life and Decline of the American Railroad*, New York, 1970, pp. 24–26.
2. Oliver Jenson, *The American Heritage History of Railroads in America*, New York, 1975, pp. 52–53.
3. Frederick T. Andrews, "The Heritage of Telegraphy," *IEEE Communications Magazine*, August 1989, pp. 12–18.
4. Ibid., p. 13.
5. Alfonso M. Albano, "Morse and the Beginnings of Telegraphy," *Episodes in American Invention*, Monograph Series of the New Liberal Arts Program, Stony Brook, N.Y., pp. 39–81. This essay appears in somewhat different form in Newton Copp and Andrew Zanella, *Discovery, Innovation, and Risk*, Cambridge, Mass., 1993, pp. 13–35.
6. Albano, op. cit., pp. 52–59. Inductance is measured with the unit of Henrys in honor of Joseph Henry.
7. Sarah R. Riedman, *Trailblazer of American Science*, New York, 1961, pp. 31–37.
8. Ibid., pp. 37–48.
9. Joseph Henry, "On the Application of the Principle of the Galvanic Multiplier to Electromagnetic Apparatus, and also to the Development of Great Magnetic Power in Soft Iron, with a Small Galvanic Element," *Silliman's American Journal of Science*, January 1831, Vol. 19, pp. 400–408. The paper is reprinted in *Scientific Writings of Joseph Henry*, Vol. 1, Washington, DC, 1886, pp. 37–49. All subsequent numbers came from this paper or are inferred from it.
10. Ibid., 1886, p. 45. Henry's 1831 paper was the first of several in which he described his experimental investigations into the phenomena of electromagnetic induction and

electrodynamics. These ideas were later put into a general theory of electromagnetism by James Clarke Maxwell.

11. Henry did not make such calculations even though he understood the relationships. Using more modern formulations, we can derive the attractive force of Henry's magnet by estimating from magnetization curves, the value of K, and then computing the force by means shown in the figures. For a standard curve see E. A. Avallone and T. Baumeister III, *Marks' Standard Handbook for Mechanical Engineers*, New York, 9th ed., 1987, p. 15–11.

12. Joseph Henry, "On a Reciprocating Motion Produced by Magnetic Attraction and Repulsion," *Silliman's American Journal of Science*, July 1831, Vol. 20, pp. 340–343. Henry, op. cit., 1886, pp. 54–57. The article does not mention the telegraph but does describe the motion produced by electromagnetism.

13. Thomas Coulson, *Joseph Henry: His Life and Work,* Princeton, 1950, p. 96. See also Herbert Bailey, "Joseph Henry," *A Princeton Companion,* Ed. Alexander Leitch, Princeton, 1978, pp. 246–249.

14. Coulson, op. cit., pp. 169–207.

15. Joseph Henry, "Statement in Relation to the History of the Electro-Magnetic Telegraph," Henry, op. cit., 1886, Vol. 2, pp. 426–436; the quote is on p. 436.

16. Brooke Hindle, *Emulation and Invention*, New York University Press, 1981, pp. 85–100.

17. Ibid., p. 101.

18. Bern Dibner, "Communication," *Technology in Western Civilization*, Vol. 1, Eds. M. Kranzberg and Carroll W. Pursell, Jr., Oxford University Press, New York, 1967, pp. 452–456.

19. Hindle, op. cit., pp. 96–98.

20. Ibid., pp. 98–100.

21. Paul J. Staiti, *Samuel F. B. Morse*, Cambridge, 1989, pp. 226–228.

22. Hindle, op. cit., pp. 105–107.

23. A. B. Cornell, *Biography of Ezra Cornell*, New York, 1884, pp. 108–119.

24. Ibid., p. 114.

25. Albano, op. cit., pp. 65–66.

8. St. Louis versus Chicago and the Continental Railroads

1. Thomas C. Clarke et al., *The American Railway: Its Construction, Development, Management, and Appliances*, New York, 1889, xxiii, xxviii.

2. Ibid., p. 1.

3. Ibid., pp. 45–46.

4. Ibid., pp. 129–131.

5. Ibid., pp. 126–127.

6. Ibid., pp. 120–121. The first locomotive given in a table on page 126 weighed 92,000 lbs. It is the one used here as an example.

7. Ibid., p. 121.

8. Ibid., p. 175 for the rates drop by 1887 and p. 344 for the wealth and investment in railraods by 1889.

9. Ibid., p. 425. The length of rail lines in the United States was less than the world's total but so many of them had double, triple, or more tracks per line that it seemed rea-

sonable by 1889 to state that the total track length exceeded that in the rest of the world.

10. Oliver Jensen, *Railroads in America*, New York, 1981, pp. 98–100.

11. John F. Stover, *American Railroads*, University of Chicago Press, 1961, pp. 67–74.

12. W. W. Belcher, *The Economic Rivalry Between St. Louis and Chicago, 1850–1880*, New York, 1947.

13. Harold M. Mayer and Richard C. Wade, *Chicago: Growth of a Metropolis*, Chicago, 1969, pp. 35–38.

14. Ibid., p. 35.

15. Annual Report of the Chief of Engineers, 1878, Vol. 2, p. 1024.

16. "Memoir of Henry Flad," American Society of Civil Engineers, *Transactions*, Vol. 42, p. 561. Flad, a graduate of the University of Munich in civil engineering, was employed on Rhine River improvements until 1849, when he fled to the United States, fearing execution for his support of the established Parliamentary government against rebellious regional princes. First associated with the Ohio and Mississippi Railroad Company, he eventually arrived in St. Louis in 1854 as the company's representative. Flad enlisted in the Union Army in 1861 as a private, but with his extensive engineering skill he had by 1864 risen to the rank of full colonel, Corps of Engineers. It was Flad who later developed the cantilever construction technique to the unprecedented extent used in the St. Louis Bridge.

17. Ralph W. Gilbert, Jr., and David P. Billington, "The Eads Bridge and Nineteenth Century River Politics," *Civil Engineering: History Heritage, and the Humanities, Background Papers*, 1970, pp. 85–116.

18. William Taussig, "Addresses Before the Missouri Historical Society," *Reminiscences by Personal Friends of General U. S. Grant*, Ed. James T. Post, St. Louis, J. L. Post, 1904, p. 144. Taussig explained that in the early stages of the bridge enterprise, the steamboat men of St. Louis opposed it, without avail, however, in the legislature, in Congress, and in the municipal assembly. They opposed the erection of any bridge whatever, ostensibly because their high chimneys could not pass under the bridge at high water, but really because they thought railroad bridges would seriously affect the river trade.

19. U.S. Congress of Representatives, *Executive Document No. 1*, Part e, 43rd Congress, 1st Session, Washington, DC, Government Printing Office, 1873.

20. William Taussig, "Addresses Before the Missouri Historical Society," *Reminiscences by Personal Friends of General U. S. Grant*, Ed. James T. Post, St. Louis, J. L. Post, 1904, p. 144.

21. C. M. Woodward, *A History of the St. Louis Bridge: Containing a Full Account of Every Step in its Construction and Erection, and Including the Theory of the Ribbed Arch and the Tests of Material*, St. Louis, 1881.

22. John A. Kouwenhoven, "The Designing of the Eads Bridge," *Technology and Culture*, Vol. 23, No. 4, October 1982, pp. 535–536.

23. Ibid., p. 568. Taussig, a physician turned banker, had been associated with Eads since about 1850. Ibid., p. 548.

24. Ibid., p. 552.

25. *Addresses and Papers of James B. Eads*, Ed. Estill McHenry, St. Louis, 1884, p. 513. This quote is given by Kouwenhoven, op. cit., pp. 544–545. Currently the longest spanning steel arch bridge is 1700 feet in span over the New River Gorge in West Virginia.

26. Woodward, op. cit., p. 351.
27. Ibid., p. 331 and plate XXIX.
28. Ibid., pp. 120–121.
29. Ibid., p. 332. The curve of the Eads Bridge is actually circular but it is such a flat arch that the difference between circular and parabolic is negligible both structurally and visually. The arch rise of the Eads Bridge at the quarterspan is 35.3 ft. whereas for a parabola with $d = 47$ ft. and $L = 520$ ft. it would be 35.0 ft.
30. Strictly speaking, the more widely the tubes are separated the more bracing steel is required between them, but this secondary steel is only a small percentage of the steel in the tubes.
31. Woodward, op. cit., p. 60.
32. Stover, op. cit., pp. 77–78.
33. Ibid., p. 82.
34. Ibid.
35. Ibid., p. 111.
36. Ibid., p. 117.
37. Ibid., pp. 126–131.
38. Ibid., p. 131.
39. Ibid., p. 136.
40. Ibid., p. 138.
41. Ibid., p. 114.
42. Leo Marx, *The Machine in the Garden: Technology and the Pastoral Ideal in America*, Oxford University Press, New York, 1964.
43. Ibid., p. 353.
44. Ibid., p. 13.
45. Ibid., p. 16.
46. *The Railroad in American Art: Representations of Technological Change*, Eds. Susan Danly and Leo Marx, MIT Press, 1988.
47. Carl Condit, *The Port of New York: A History of the Rail and Terminal System from the Beginnings to Pennsylvania Station*, Chicago, 1980, pp. 83–93 and 122.
48. Ibid., p. 105.

9. Carnegie and the Climax of Steel

1. Harold C. Livesay, *Andrew Carnegie and the Rise of Big Business*, Boston, 1975, pp. 1–22.
2. Ibid., pp. 23–42.
3. Ibid., pp. 45–46, 53–54.
4. Ibid., p. 71.
5. James A. Ward, *J. Edgar Thomson*, op. cit., pp. 130–131.
6. T. C. Clarke, *The American Railway*, op. cit., p. 37. Clarke states that steel rails last five or six times as long as iron ones, while Ward, op. cit., states that they lasted about eight times as long.
7. J. E. Gordon, *The New Science of Strong Materials: or Why Don't You Fall Through the Floor*, 2nd ed., Princeton, 1976, p. 44.
8. W. C. Roberts-Austen, *Metallurgy*, London, 1892, p. 92.
9. Gordon, op. cit., p. 221, describes how scientific theory came after "most of the possible important improvements have already been made by traditional and empirical methods."

10. Frank T. Sisco, *Modern Metallurgy for Engineers*, 2nd ed., New York, 1948, pp. 36–40.
11. *Encyclopedia Britannica*, Vol. 17, 1963, p. 921.
12. R. A. Mott, *Henry Cort: The Great Finer: Creator of Puddled Iron*, The Metals Society, London, 1983.
13. Gordon, op. cit., p. 243.
14. Jeanne McHugh, *Alexander Holley and the Makers of Steel*, Baltimore, 1980, pp. 71–75.
15. Henry Bessemer, *Sir Henry Bessemer: An Autobiography, 1905*, p. 93. This work was unpublished and unfinished at Bessemer's death in 1898. See note 1, p. 76, of McHugh, op. cit.
16. Jacques Barzun, "The Paradoxes of Creativity," *The American Scholar*, Summer 1989, p. 347.
17. Robert C. Binkley, *Realism and Nationalism: 1852–1871*, New York, 1935, p. 173.
18. McHugh, op. cit., p. 78.
19. Ibid., p. 88.
20. W. T. Jeans, *The Creators of the Age of Steel*, London, 1884, pp. 39–43.
21. McHugh, op. cit., p. 171.
22. Alexander L. Holley, *American and European Railway Practice in the Economical Generation of Steam*, New York and London, 1861, 192 pages with 77 plates. The quotation is about boilers and appears on p. 29.
23. Quoted in McHugh, op. cit., pp. 369–370.
24. Ibid., pp. 1–42.
25. Ibid., p. 52.
26. Zerah Colburn and Alexander L. Holley, *The Permanent Way and Coal-Burning Locomotives of European Railways: With a Comparison of the Working Economy of European and American Lines, and the Principles upon which Improvement Must Proceed*, New York, 1858, 168 pages with 51 plates.
27. McHugh, op. cit., pp. 61–77, 168, and 170–171.
28. John N. Ingham, *Making Iron and Steel: Independent Mills in Pittsburgh, 1820–1920*, Columbus, Ohio, 1991, pp. 205–206.
29. Elting E. Morison, *Men, Machines, and Modern Times*, Cambridge, Mass., 1966, pp. 184–185.
30. Ibid., p. 189. Morison paraphrases this comment but does not give a source for it.
31. Livesay, op. cit., p. 114.
32. Joseph Frazier Wall, *Andrew Carnegie*, Pittsburgh, 1970, pp. 307–310.
33. Ibid., pp. 311–316.
34. Livesay, op. cit., p. 101.
35. Ibid., p. 117.

10. Edison and the Network for Light

1. Harold C. Passer, *The Electrical Manufacturers: 1875–1900*, Harvard University Press, 1953, pp. 78–79.
2. Thomas B. Hughes, "The Electrification of America: The System Builders," *Technology and Culture*, Vol. 20, No. 1, January 1979, p. 126. The following numerical discussion is derived in part from Hughes, ibid., pp. 135–138.
3. Charles L. Clarke, quoted in Francis Jehl, *Menlo Park Reminiscences*, Vol. 2, Dearborn, Mich., 1938, pp. 852–854.

4. Hughes, op. cit., p. 134.

5. Thomas B. Hughes, "Thomas Alva Edison and the Rise of Electricity," *Technology in America: A History of Individuals and Ideas*, MIT Press, 2nd ed., 1990, Ed. Carroll W. Pursell, Jr., p. 124.

6. Jehl, op. cit., Vol. 2, pp. 568, 570.

7. Bernard Gorowitz, Ed., *A Century of Progress: The General Electric Story*, Schenectady, N.Y., 1981, Vol. I, p. 11.

8. Jehl, op. cit., Vol. 3, p. 1050. Edison bought Porter-Allen engines, which had already proven superior to Corliss ones for high speed and relatively even performance (to prevent flickering lights). Even these, however, turned out to be unsatisfactory and were replaced by another make and modified in their controls by Edison.

9. Ibid., Vol. 1, pp. 374–376.

10. Ibid., Vol. 2, p. 650.

11. Values are taken from Thomas B. Hughes, *Networks of Power: Electrification in Western Society, 1880–1930,* Johns Hopkins University Press, Baltimore, Md., 1983, p. 39.

12. George Wise, "Swan's Way: Inventive Style and the Emergence of the Incandescent Lamp," *IEEE Spectrum*, April 1982, pp. 66–70.

13. Hughes, op. cit., 1983, p. 30.

14. John W. Winkler, *Morgan the Magnificent*, New York, 1930, pp. 75–76.

15. Hughes, op. cit., 1983, p. 39.

16. Ibid., p. 46.

17. Ibid., p. 42.

18. Ibid., pp. 25–29. Here Hughes gives a balanced argument for Edison's greatness, and my discussion is largely based upon that.

19. Ibid., p. 18.

20. Thomas Coulson, *Joseph Henry, His Life and Works*, Princeton University Press, 1950.

21. Robert Silverberg, *Light for the World: Edison and the Power Industry*, Van Nostrand, Princeton, N.J., 1967, pp. 229–243.

22. Ibid., pp. 238–239. See also Hughes, op. cit., 1983, pp. 112–117.

23. Henry Adams, *The Education of Henry Adams*, Modern Library Inc., New York, 1931 (originally published by the Massachusetts Historical Society in 1918), pp. 342–343.

24. This and the following comes directly from Robert Prigo, *Nikola Tesla and the A.C. Motor*, draft for a monograph for the New Liberal Arts Program, 1989, pp. 7–9.

25. Margaret Cheney, *Tesla: Man Out of Time*, Englewood Cliffs, N.J., Prentice Hall, 1981, pp. 21–26.

26. Ibid., p. 130.

27. A.J. Beckhard, *Nikola Tesla, Electrical Genius*, London, Dodson, 1961, pp. 116–119.

28. Ibid., p. 130.

29. John J. O'Neill, *Prodigal Genius: The Life of Nikola Tesla*, New York, Washburn, 1944, p. 108.

30. Chaney, op. cit., pp. 40, 48–49

31. Ronald Kline, "Science and Engineering Theory in the Invention and Development of the Induction Motor, 1880–1900," *Technology and Culture*, 1987, pp. 287–291.

11. The Centennial Revolutions, 1876–1883

1. Nash et al., *The American People*, 1986, does not include it at all; Henrietta et al., *America's History*, 1987, does not discuss the centennial except in the picture caption

for the Corliss engine on p. 520. Link et al., *The American People*, 1981, does give a brief paragraph on it on p. 494, and Blum et al., *The National Experience*, 1985, treats it disparagingly in a brief section on p. 412.

2. Both quotations come from *1876: A Centennial Exhibition*, Ed. Robert C. Post, Washington, DC, 1976, pp. 8 and 9.

3. Phillip T. Sandhurst, *The Great Centennial Exhibition 1876*, P. W. Ziegler & Co., Philadelphia, pp. 361–369.

4. Mary Black, *Old New York in Early Photographs: 1853–1901*, New York, 1973, pp. vi–x.

5. Dian O. Belanger, "The Corliss at Pullman," *Technology and Culture*, Vol. 25, No. 1, January 1984, pp. 83–90.

6. David McCullough,*The Great Bridge*, New York, 1972.

7. Alan Trachtenberg, *Brooklyn Bridge: Fact and Symbol*, New York, 1965.

8. Belanger, op. cit., p. 85.

9. Stanley Buder, *Pullman: An Experiment in Industrial Order and Community Planning, 1880–1930*, New York, 1967.

10. David P. Billington, *The Tower and the Bridge*, Chapter 5.

11. J. A. Roebling, "Report and Plan for a Wire Suspension Bridge," *Order of Preference of the Supreme Court of the United States*, Saratoga Springs, N.Y., 1851, p. 479.

12. J. A. Roebling, "Memoir of the Niagara Falls Suspension and Niagara Falls International Bridge," *Papers and Practical Illustrations of Public Works*, London, 1856, pp. 4–5.

13. Ibid.

14. Trachtenberg, op. cit., pp. 101–109.

15. J. A. Roebling, "Report of John A. Roebling, Civil Engineer, to the President and Directors of the New York Bridge Company on Proposed East River Bridge," Trenton, N.J., September 1, 1867, pp. 17–19 (published in Brooklyn, N.Y., 1870).

16. Montgomery Schuyler, "The Bridge as a Monument," *Harper's Weekly*, 27, May 1883, p. 326; reprinted in *American Architecture and Other Writings*, Eds. William H. Jordy and Ralph Coe, Atheneum, New York, 1964, p. 164.

17. Ibid., p. 173.

18. Ibid., p. 173.

19. Sandhurst, op. cit., pp. 361–363.

20. Belanger, op. cit., p. 84.

21. John F. Kasson, *Civilizing the Machine*, Penguin, 1977, p. 183.

22. John A. Garraty, Ed., *The Transformation of American Society, 1870–1890*, Harper & Row, New York, 1968, p. 2.

23. Ibid., p. 4.

24. Woodrow Wilson, "The House of Representatives," 1885, reprinted in Garraty, op. cit., pp. 202–210.

25. Eugene S. Ferguson, "Power and Influence: The Corliss Steam Engine in the Centennial Era," *Bridge to the Future: A Centennial Celebration of the Brooklyn Bridge*, Eds. M. Latimer, B. Hindle, and M. Kranzberg, *Annals of the New York Academy of Sciences*, Vol. 424, 1984, p. 244.

26. Quoted in ibid., pp. 225–228.

27. Kasson, op. cit., p. 244.

28. Ferguson, op. cit., p. 244.

29. For its competitiveness, see John T. Henthorn and Charles D. Thurber, *The Corliss Engine [and] Its Management*, New York & London, 1897, 92 pages plus an appendix. This little book begins as follows:

> Since the issue of the patent, in the year 1849, to Geo. H. Corliss, for certain improvements in the working of steam engines, and covering the admission of steam to the cylinder by the combined action of a governor, to determine the point of cutoff at which a liberating valve-gear shall act, and thus allow a certain amount of expansion to take place in the cylinder before the end of a stroke is completed, I think it will be conceded by all fair-minded engineers, when we come to look over the ground carefully, covered by the proposition, that no improvement has been made since that time, up to this date, in the economy of working steam expansively as exemplified by this system.

30. Henry Adams, *The Education of Henry Adams*, Ed. Ernest Samuels, Boston, 1976, p. 380.
31. Irving K. Pond, "Pullman—America's First Planned Industrial Town, by a Collaborator and Eyewitness," Illinois Society of Architects *Monthly Bulletin*, 18–19, June–July 1934, p. 7.
32. Belanger, op. cit., p. 86.
33. Belanger, op. cit., p. 84.
34. Andrew Carnegie, *Triumphant Democracy: Sixty Years' March of the Republic*, New York, 1886, p. 509. This section is quoted in John A. Garraty, Ed., *The Transformation of American Society, 1870–1890*, New York, 1968, p. 25.
35. Garraty, op. cit., p. 22.
36. See ibid., Part VII, "Social Reformers," pp. 228–265.

Index

Adams, Charles Francis Jr., 140
Adams, Henry, 140, 196, 217
Aetna steamboat, 55–58
Albany Academy, 122
alternating current. *See* electric power network
American and European Railway Practice (Holley), 172
American Passenger Locomotive, 141
American Society of Civil Engineers, 93, 94
American Society of Mechanical Engineers, 63
Ammann, Othmar, 12
Ampère, André Marie, 121
Anderson, John, 23
Appleton, Nathan, 69, 72, 74, 76
Appleton Mills, 83
arc light, 187
arch bridges, 6–8, 31–34, 145–54
Arthur, Chester A., 154, 156

Baldwin, James F., 79
Baldwin, Matthias, 109–12
    Lancaster locomotive, 110–12, 162
Baldwin Locomotive Works, 140
Baltimore and Ohio Railroad, 115
Barlow, Peter, 94
Barnum, P. T., 188
Batchelor, Charles, 192
battery. *See under* Volta, Alessandro
Bayway Refinery, 186
Beaufoy, Mark, 45, 53
    water drag test results, 45, 49, 51, 175
Beck, T. Romeyn, 122, 123
Belknap, General William W., 146–47

Berlin, Isaiah, 192
Berlioz, Hector, 169
Berman, Solon S., 217
Bessemer, Henry, 168–72
    developed Bessemer steel smelting process, 162, 169–72
    See also Holley, Alexander Lyman
Bessemer converters, 14, 170, 173, 175
blast furnace, 17–19, 165–68
bogie truck. *See* swivel truck
boilers. *See under* Evans, Oliver
Bonaparte, Napoleon, 42, 139
Bonar Bridge, 32–34
Boomer, L.B., 146
Bossut, Charles, 53
Boston Associates, 74, 79
Boulton, Matthew, 26–29, 34, 37, 38
    letter to James Watt, 37
    partnership essential to Watt, 26–29
Boulton and Watt Company, 12, 101
Boyden, Uriah, 83, 90, 93
Boyden turbine, 83
    See also Francis turbine
Braithwaite and Ericson Company, 100
Bridgewater Canal, 98
Bristol Floating Harbour, 107
Britannia Bridge, 105, 108
    compared to Saltash Bridge, 108
British Association for the Advancement of Science, 172
British Parliament, 26, 54, 100
broad gauge. *See* railway gauges
Brooklyn Bridge, 37, 154, 162, 163, 199–201, 204, 206, 207–12, 218, 219
    cost and politics, 209–12
    design by John Roebling, 207–9

Brooklyn Bridge (*continued*)
  symbol of late-nineteenth-century
     America, 210–12
  See also Roebling, John
Brown, A. K., 197
Brown, Gratz, 145
Brown University, 174
Brunel, Isambard Kingdom, 105–8, 193
  Clifton Gorge Bridge, 107
  Bristol Floating Harbour, 107
  gauge war with Robert Stephenson,
     107–8
  Great Western Railway, 107
  Saltash Bridge, 108
  Sultan locomotive, 108
Brunel, Marc, 105
  Thames tunnel, 105
Brush, Charles F., 187
Buildwas Bridge, 31–32
Bush, Vannevar, 52

cable bridges, 7–10
Caledonian Canal, 30
Calhoun, John C., 74
Camden and Amboy Railroad, 115
canals (for industrial water power)
  See Francis, James B.; Northern Canal
Carnegie, Andrew, 19, 117, 149, 156, 158,
     159–163, 175–77, 179–80, 219
  bridge-building, 161
  compared to Thomas Edison, 179–80
  early life and wealth, 159–62
  founding of steel industry, 162–63,
     175–77
  idea of wealth as a trust, 219
  See also Holley, Alexander Lyman; steel
Carnegie, Will, 159
Carnegie, McCandless and Company, 175,
     177
Carnot, N. L. Sadi, 53
Centennial Exhibition. *See* Philadelphia
     Centennial Exhibition
Central Pacific Railroad, 142
Cézanne, Paul, 39
Chicago, 206, 207
  rivalry with St. Louis, 143–44
Chicago Columbian Exposition (1893),
     37, 196, 217
Clarke, Thomas Curtis, 140

Clemens, Samuel (Mark Twain), 63
Clermont steamboat, 43–44, 55
Cleveland, Grover, 157
Clifton Gorge Bridge, 107
Colburn, Zerah, 174
Cole, Thomas, 4, 14
College of New Jersey (later Princeton
     University), 128, 131, 181, 194, 195
Columbia College (later Columbia Univer-
     sity), 198
Columbian Exposition. *See* Chicago Co-
     lumbian Exposition
Comet steamboat, 60
Cooley, Thomas M., 139, 157
Corliss, George H., 174, 176, 212–17
  design of Corliss engine, 213–14,
     215–16
Corliss engine, 37, 199–201, 204, 206–
     207, 213–18
  centerpiece of 1876 Philadelphia Cen-
     tennial Exhibition, 199–201
  powered New York City rapid transit,
     217
  powered Pullman, Illinois, 217–18
  symbol of late-nineteenth-century
     America, 214–17
Cornell, Ezra, 119, 134–36
  endowed Cornell University, 136–37
  wired telegraph network for Morse,
     134, 135
Cornell University, 136–37
Cort, Henry, 168
cow catcher, 109
Craigellachie Bridge, 33–34, 38
Crane, Hart, 206
Crimean War, 169
Crocker, Charles, 142
Cullom, Shelby M., 156–57

Daguerre, Louis, 134
Darby, Abraham, I, 22
Darby, Abraham, III, 22, 31
Davis, John, 62
Davis, John C., 109
Delaware and Hudson Canal Company,
     108
Despatch steamboat, 60–1
direct current. *See* electric power network
Dodge, Grenville M., 142, 143

Dom Pedro (emperor of Brazil), 199
Dredge, James, 173
Drexel, Morgan and Company, 189
Drew, Daniel, 157
Dripps, Isaac, 110
Durant, Thomas C., 143
dynamo (electric), 188, 190–91, 196
    in Edison's Pearl Street power station, 190–91
    symbol of industry to Henry Adams, 196

Eads, James B., 143–54, 176
    cost and politics of St. Louis Bridge (Eads Bridge), 144–49
    design of St. Louis Bridge, 149–54
Eads Bridge, 143–54, 162
Edison, Thomas A., 114, 131, 169, 179–94, 196–97
    compared to Andrew Carnegie, 179–80
    design for incandescent lightbulb, 180–86
    integration of technology and business, 189
    laboratory at Menlo Park, N. J., 180–81
    Pearl Street power plant and network, 180, 186–93
    conflict with Westinghouse and Tesla, 192–96
    See also electric power network; incandescent light
Edison Companies, 186, 189, 197, 198
Eiffel Tower, 37, 39
Einstein, Albert, 39
electric battery. See Volta, Alessandro
electric circuit, 120–21, 127
    See also telegraph; electromagnetism
electric light. See arc light; incandescent light
electric power network
    Edison and direct current, 192–93
    Pearl Street power plant and network, 186–91
    Westinghouse, Tesla, and alternating current, 193–98
    See also transformer
electromagnetism, 3–4, 123–27
    See also electric circuit; Henry, Joseph

Ellesmere Canal and Company, 29–30
Ellet, Charles, 209
engineering. See modern engineering
Engineering Founders Societies, 54
Enterprise steamboat, 60
Ericson, John, 100
Erie Canal, 97, 108, 113, 115, 117, 218
Evans, Oliver, 55, 57
    Aetna steamboat, boiler stresses in, 55–58

Faraday, Michael, 180, 194
Ferraris, Galileo, 198
Fillmore, Millard, 119
Fitch, John, 41–42, 55
Fitch-Rumsey steamboat competition, 41–42
Flad, Henry, 146
formulas, 3–4, 6–19
    networks and processes, 204–5
    structures and machines, 202–3
Francis, James B., 78, 79–94, 148, 169, 176, 180, 188
    early water management studies at Lowell, 79–81
    full-scale hydraulics testing laboratory created, 81–83
    redesigned water turbine as part of industrial water power network, 83–89
    theory and practice in approach of, 94
    water flow (weir) tests, 89–93
Francis turbine, 83–89, 162, 180
Franklin, Benjamin, 41, 135
Franklin Institute, 62
French Revolution, 69
Ford, Henry, 114
Forney, M. N., 140
Fulton, Robert, 21, 39, 42–51, 58–59, 97, 101, 132, 148, 169, 175, 176, 180, 186, 192, 193, 218
    artist as engineer, 42, 44–45, 169
    early French steamboat, 42, 53
    innovation exemplified by, 52–55
    New Orleans steamboat, 59
    North River (later Clermont) steamboat, 43–44
    partnership with Robert Livingston, 42–43

Fulton, Robert (*continued*)
  patent design for steamboat, 45–51
  rivalry of Fulton group with Henry
    Shreve on the Mississippi,
    59–62
  Vesuvius steamboat, 59
Fulton steamboat, 42–51, 53, 113, 180

Gale, Leonard, 3–4, 134, 135
  scientific analysis of Morse's telegraph,
    134
  See also telegraph; Morse, Samuel F. B.
Galton Bridge, 34
Galvanic Multiplier (Henry), 147
George Washington Bridge, 12
Gibb, Sir Alexander, 38
Gilbert, David, 35
Gold and Silver Telegraph Company,
  180
Golden Gate Bridge, 12
Gooch, Daniel, 107
Gould, George J., 157
Gould, Jay, 117, 157, 160
Gramme, Zénobe, 187
Grant, Ulysses S., 143, 147, 156, 199
Graz Polytechnic, 197
Great Western Railway, 107

Haigh, J. Lloyd, 210
Harriman, Edward H., 114, 117, 157
Harvard University, 68
Hawthorne, Nathaniel, 158
Hazledine, William, 33
The Hedgehog and the Fox (Berlin),
  192
Henry, Joseph, 122–29, 133, 134, 135,
    147, 148, 194–95, 210
  early career, 122–23
  electromagnetic research, 121,
    123–27
  electric telegraph, 126, 128–29
  role in planning Smithsonian Institu-
    tion, 128
  work at Princeton, 128–31
  See also electric circuit; Morse, Samuel
    F. B.; telegraph
Hill, James J., 114
History of the St. Louis Bridge (Wood-
  ward), 147–48

Holley, Alexander Lyman, 172–75, 176,
  177
  brought Bessemer steel process to the
    United States, 172–73
  early work in railroads and steel, 174
  collaboration with Andrew Carnegie,
    175
  See also Bessemer process; Carnegie,
    Andrew; steel
Hopkins, Mark, 142
Hopper, Edward, 158
horsepower
  Watt's formula for, 47, 55, 213
Horseshoe Curve, 117
Howells, William Dean, 216
Hughes, Thomas, 192
Humber Bridge, 209
Humphreys, General A., 147
Hunter, Louis, 53–54
Huntington, Collis P., 142
hydraulics. *See* Francis, James B.; Francis
  turbine

Illinois Central Railroad, 217
incandescent light, 16, 180–86, 188
  Edison's design for filament lightbulb,
    180–86
  Edison's Pearl Street power plant and
    network, 190–93
  See also Edison, Thomas
Industrial Revolution, 22, 37–38, 162–63
  two sides of, structures and machines,
    35
Inness, George, 158
innovation
  role of individual genius in, 54–55,
    189–92
  role of science in, 52–53, 176, 187–88
  role of social processes in, 53–54, 175–
    76, 188–89
Interborough Rapid Transit (New York
  City), 217
Interstate Commerce Commission, 62,
  139, 157
Iron Bridge, 22, 31, 163, 164
iron, industrialized, 17–19, 22, 31, 162–68
  puddling process, 168
  smelting process, 17–19, 22, 164–68
  See also steel

J. Edgar Thomson Steel Works, 175, 176
Jackson, Andrew, 135
Jackson, Patrick, 74, 76, 78
James Family, 122
James, Henry Jr., 122, 158
Jefferson, Thomas, 41, 69
Jehl, Francis, 192
Jervis, John B., 109
    pioneered American locomotive design,
        110
Johnson, Andrew, 142
Jones, Captain Bill, 175, 176, 177
Joule's Law, 15–16, 181–183, 193

Kemmler, William, 196
Kendall, Amos, 135
Keokuk Northern Line Packet Company,
    146
Kepler, Johannes, 39
Keystone Bridge Company, 161
Knox, Henry, 41
Kreusi, John, 192

Lackawanna Valley (Inness), 158
Lancaster locomotive, 110–12
Life on the Mississippi (Twain), 65
Linden Power Plant, 186
Liverpool and Manchester Railway,
    98–100
Livingston, Robert, 42–43, 59
    partnership with Robert Fulton, 42–43
    See also Fulton, Robert
Locks and Canal Company, 76, 79, 81
locomotives (steam)
    American and British compared, 109
    symbol of nineteenth century America,
        157–58
    See also under Trevithick, Richard;
        Stephenson, George and Robert;
        and Baldwin, Matthias
London and Birmingham Railway, 105
Longdon Aqueduct, 31
Louisiana Purchase, 59
Lowell, Francis Cabot, 21, 67–74, 176
    borrowed English textile technology
        and Scottish idea of planned com-
        munity, 67, 69
    contrast of Lowell and Robert Owen,
        70

founded American textile manufactur-
    ing industry, 71–74
ideal of better industrial design without
    bad social conditions, 69–70
Lowell Family, 68
Lowell, Massachusetts, 67, 79, 175
    first American industrial city, 74–78
    See also Francis, James B.
Lowell Hydraulic Experiments (Francis),
    83–93, 147
Lowrey, Grosvenor P., 189

machines, 6, 12–14, 19–20, 201–4
Magnetic Telegraph Company, 135
magnetism. *See* electromagnetism
Marx, Leo, 157–58
McCandless, David, 176
McCormick, Cyrus Hall, 192
McCune, John S., 146
Men of Progress (Schussele), 135
Menai Straits Bridge, 30, 32, 35–36, 201
Menlo Park, N.J.
    Edison's laboratory at, 180–81
Merriam, John C., 62
Merrimack Manufacturing Company, 67,
    76, 78
Michigan Central Railroad, 217
Middlesex Canal, 76
Minot, Charles, 119–20
modern engineering
    basic innovations, vii
    efficiency, economy, and elegance,
        goals of, 39, 154, 204–6
    four great ideas of, 6–20, 201–4.
        See also structures, machines, net-
            works, processes
    historical debate over, 37–39
    interconnectedness of, 19–20, 201–6
    liberal arts and engineering, 5–6
    objects and systems, 206–7
    science and engineering, 11, 35, 176,
        211
    stimulus to social ideas, 37–39, 69–70
    three perspectives on, vii, 204–6.
        See also scientific, social, and
            symbolic perspectives
    See also innovation
Mohawk and Hudson Railroad, 113
Monitor steam gunboat, 100

Monroe, James, 59
Moody, Paul, 71–72, 74, 76
Morgan, J. Pierpont, 114, 156, 160, 189
Morrill Act (1862), 136
Morse, Samuel F. B., 3–4, 128, 129, 131–
    36, 169, 176, 186, 189, 192
    artistic career, 132, 169
    disagreement with Joseph Henry, 133
    formed Western Union Company, 133
    founding of American telegraph net-
        work, 133–36
    partners of, 134–35
    See also telegraph; Henry, Joseph
Morse code, 133, 135

narrow gauge. See railway gauges
National Academy of Design, 132
National City Bank, 157
National Portrait Gallery, 135
National Science Foundation, 52, 54
Navier, C.L.M.H., 36, 94
networks, 6, 14–17, 19–20, 201–5
Networks of Power (Hughes), 192
Newcomen, Thomas, 22, 23–24, 193
    design of atmospheric steam engine,
        23–26
Newcomen engine, 22, 23–26, 31, 131
    improved by Smeaton, 25
New Jersey Turnpike, 186
New Orleans steamboat, 59
New River Gorge Bridge, 34
Newton, Sir Isaac, 39
New York and Erie Railroad, 119
New York Central Railroad, 113, 155–56
New York, Providence, and Boston Rail-
    road, 79
New York State Legislature, 42
New York University, 132, 133
North Star locomotive, 107
Northern Canal, Lowell, 81–83
Northern Pacific Railroad, 154
Northern Railway, 154
Northumbrian locomotive, 100

Oersted, Hans Christian, 120, 121
Ohm, Georg Simon, 121
Ohm's Law, 15–16, 121, 181–83, 193,
    202
On Growth and Form (Thompson), 19

Ordnance and Armor (Holley), 174
O'Reilly telegraph office, 159
Owen, Robert, 70

Paris Expositions (1889, 1900), 37–38
Patent Application (Fulton), 45–51, 147
patents and patent law
    Edison's patents, 192
    Evans steamboat patent, 55
    first U.S. patent law, 41
    Fulton steamboat patent, 45–51
    patent protection and government regu-
        lation, 58
    Watt separate condenser patent, 12, 26
Patrick locomotive, 78
Pearl Street power plant, 186–92, 193
    See also Edison, Thomas
Pendaron Iron Works, 97
Pennsylvania Railroad, 113–17, 119, 156,
    159, 161, 162
    See also Thomson, J. Edgar
Philadelphia and Columbia Railroad, 108,
    110
Philadelphia Centennial Exhibition
    (1876), 199, 206, 212–13, 218, 219
    See also Corliss engine
photography, 134
Piper, John, 161
Pont-y-Cysyllte Aqueduct, 31
Portage Railroad, 117
Port of New York Authority, 6
Prague University, 197
Princeton University. See College of New
    Jersey
Pritchard, Thomas, 22, 31
processes, 6, 17–20, 162–72, 201–5
Promontory Point, 143, 154
Prony dynamometer, 83, 85–87
puddling process. See under iron, industri-
    alized
Pullman, George, 217–18
    planned city of Pullman, Illinois, 207,
        217–18

railroads, American
    compared to British, 109
    early lines, 108–9
    innovations by Baldwin and others,
        109–12

integration of technology and business by J. Edgar Thomson, 113–18
power and extent by 1880s, 140–142
railroads, British
　origins, 98
　Rainhill trials, 98–100, 108
　Stephenson designs, 98–100, 101–5
railway gauges, 105–8
　Brunel broad gauge, 107–8
　Stephenson narrow gauge and its adoption, 100, 107–8
Railway Standard Time, 137
Rainhill trials, 99–100, 108
Randolph, Edmond, 41
Reagan, John H., 156
regulation by government, 218
　railroads, 156–57
　steamboats, 58–59, 61–63
Rennie, George, 172
Rensselaer Polytechnic Institute, 122
Renwick, James, 128
Reynolds, William, 32
Robert Stephenson and Company, 99, 107
Roberts, W. Milnor, 148–49
Rockefeller, John D., I, 19, 117, 160
Rocket locomotive, 99
Roebling, John A., 108, 149, 207–12, 217
　cost and politics of Brooklyn Bridge, 209–12
　design for Brooklyn Bridge, 207–9
　See also Brooklyn Bridge; Schuyler, Montgomery
Roebling, Washington, 210
Roebling Company, 210
Roosevelt, Theodore, 157
Rumsey, James, 41–42, 55

St. Louis, 143–44
St. Louis and Illinois Bridge Company, 145
St. Louis Board of Water Commissioners, 146
Saint-Saëns, Camille, 169
Saltash Bridge, 108
Santa Fe Railroad, 154
Schussele, Christian, 135
Schuyler, Montgomery,
　criticism of Brooklyn Bridge, 211–12

scientific perspective, vii, 3–4, 6–7, 12–13, 14–19, 201–5
　bridges, 6–7, 31–36, 147–54, 207–9
　electric power network, 190–91, 192–196
　incandescent light, 15–17, 181–88
　industrialized iron, 17–19, 22, 31, 162–68
　railroads, 98–99, 101–5, 109–12
　steamboats, 41, 45–51, 55–58
　steam engines, 12–14, 23–26, 212–14, 215–16
　steel, 163, 168–172
　telegraph, 15, 17, 120–29, 134
　water power network, 71, 74–78, 79–93
Scott, Thomas A., 159–160
Sequin, M., 102
Sheeler, Charles, 158
Shields, James, 62
Shreve, Henry Miller, 58–61
　Comet steamboat, 60
　Despatch steamboat, 60, 61
　Enterprise steamboat, 60–61
　rivalry with Fulton group, 59–62
　Washington steamboat, 61
Siemens, Werner, 187
Silliman, Benjamin, 128
Smeaton, John
　improved Newcomen engine, 25
Smith, Adam, 22
Smithson, James, 128
social perspective, vii, 4–5, 7–9. 13–14, 17, 19, 205–6
　bridges, 7–9, 11, 144–47, 154, 209–10
　electric power network, 188–89, 192–93, 196
　incandescent light, 17, 188–89
　industrialized iron, 22, 162–63
　innovators' need for social and technical infrastructure, 175–76
　railroads, 107–8, 109–10, 113–18, 139–44, 155–58
　steamboats, 58–63, 144
　steam engines, 13–14, 37–38, 214–17
　steel, 162–63, 172, 175, 176–77
　telegraph, 119–20, 132–37
　water power network, 69–74, 93
Southern, John, 32
Southern Pacific Railroad, 154

Stanford, Leland, 142, 143, 154
steamboat. *See* Evans, Oliver; Fitch, John;
    Fulton, Robert; Fulton steamboat;
    Rumsey, James; Shreve, Henry
steam engine. *See* Corliss engine; New-
    comen engine; Watt engine
steam locomotive. *See* Baldwin, Matthias;
    Stephenson, George; Stephenson,
    Robert; Trevithick, Richard
steel
    superiority to iron, 163
    Bessemer process for smelting, 168–72
    See also Bessemer, Henry; Carnegie,
        Andrew; Holley, Alexander Lyman
Stella, Joseph, 206
Stephenson, George, 98–100, 104, 140
    first locomotive engine and railroad
        line, 98
    Rainhill trials, 98–100, 108
    Rocket locomotive, 99
    Northumbrian locomotive, 100
Stephenson, Robert, 78, 98, 140, 180,
    186
    Britannia Bridge, 105, 108
    mature locomotive design, 101–5
    gauge war with Brunel, 107–8
    North Star locomotive, 107
    Stourbridge Lion locomotive, 108
Stevens, Edwin, 174
Stillman, James, 157
Stockton and Darlington Railway, 98
Storrow, Charles S., 79
Stourbridge Lion locomotive, 108
structures, 6–12, 19–20, 199–207
    as works of art, 33–34, 39, 149, 210–12
Sultan locomotive, 108
swivel truck, 109
symbolic perspective, vii, 4–5, 9–12, 14,
    17, 19, 206
    bridges, 9–12, 38–39, 154, 199–201,
        206, 210–12
    electric light, 189–192
    railroads, 4–5, 109, 147, 157–58
    steamboats, 63–65
    steam engines, 4, 14, 206, 214–17
    telegraph, 135

Tariff of 1816, 74
Taussig, William, 147, 149

telegraph (electric)
    early telegraphy, 120
    Henry telegraph, 128–29
    Morse and the telegraph network,
        131–35
    Morse code, 133
    railway signalling, 119–20
    U.S. telegraph network in 1884, 137
    See also electric circuit; electromagnet-
        ism; Henry, Joseph; Morse,
        Samuel F. B.
Telford, Thomas, 21, 22, 29–39, 69, 149,
    207
    artistic approach to design, 38–39
    Bonar Bridge, 32–33, 34
    Buildwas Bridge, 31–32
    Caledonian Canal, 30
    compared to James Watt, 34–35
    construction methods, contrast to manu-
        facturing, 35
    Craigellachie Bridge, 33–34, 38
    Ellesmere Canal, 29, 30–31
    Galton Bridge, 34
    Longdon Aqueduct, 31
    Menai Straits Bridge, 30, 32, 35–36
    Pont-y-Cysyllte Aqueduct at Llan-
        gollen, 31
    Thames River bridge (unbuilt), 32
    Tewkesbury Bridge, 34
Tennessee Valley Authority, 6, 193
Tesla, Nikola, 192, 197–98
    conceived alternating current motor,
        197–198
    disagreement with Thomas Edison, 197
    work with George Westinghouse, 198
Tesla Electric Company, 197
Tesla motor, 198
textile industry. *See* Lowell, Francis; Fran-
    cis, James B.; Lowell, Massachusetts
Thames River bridge, 32
Thames tunnel. *See under* Brunel, Marc
thermodynamics
    stimulated by steam engine, 35
Thomson, Elihu, 195
Thomson, J. Edgar, vii, 113–18, 119, 155,
    159, 160, 162, 176, 177
    early railroad work, 114–116
    builder of the Pennsylvania Railroad,
        114–18

integration of technology and business, 113–15
  personal qualities of, 117–18
Thomson Steel Works. *See* J. Edgar Thomson Steel Works
transformer (electric), 194–96
  safety value of, 195
Treatise on Canal Navigation (Fulton), 42
Trevithick, Richard
  early locomotive engine of, 97–98
Truman, Harry, 52
turbine. *See* Boyden turbine; Francis turbine
Turner, J. M. W., 39
Twain, Mark. *See* Clemens, Samuel

Union Pacific Railroad, 140, 142, 162
U.S. Army Corps of Engineers, 146–47, 149
U.S. Congress, 54, 55, 62–63, 134–35, 142–46, 156–57, 215, 219
  authorization of Eads bridge, 145–46
  interest in telegraph, 134–35
  regulation of railroads, 156–57
  regulation of steamboats, 62–63
United States
  transformation by modern engineering, 4–6, 214–15, 218–20
Upton, Francis R., 181–85, 187, 189, 192
  analysis of incandescent light for Thomas Edison, 183–85

Vail, Alfred, 134, 135
Van Buren, Martin, 134
Vanderbilt, Cornelius, 113, 117, 155–56, 157, 158, 160
  rivalry with Pennsylvania Railroad, 117
Vanderbilt, William H., 156
Van Rensselaer, General Stephen, 122
Vesuvius steamboat, 59
Volta, Alessandro, 120–21
  voltaic battery of, 120

Waltham, Mass., 71, 74, 76
War of 1812, 70, 74
Washington, George, 41
Washington steamboat, 61
Watt, James, 12, 21, 22, 23–29, 31, 42, 47, 53–54, 55, 59, 101, 180, 186, 193
  added separate condenser to Newcomen engine, 13, 26
  formula for horsepower, 25
  need for Matthew Boulton as partner, 26–29
  other inventions, 26–27
  studied Newcomen steam engine, 23–26
  Watt and Thomas Telford compared, 34–38
  See also patents and patent law
Watt engine, 25–26, 31, 38, 42, 45, 47, 55, 131, 175, 180, 193
  in Fulton steamboat, 45, 175
  rotary engine, 27
  See also Watt, James
Wealth of Nations (Smith), 22
Webster, Daniel, 119
weir studies. *See under* Francis, James B.
West, Benjamin, 42, 132
Western Union Telegraph Company, 133, 136, 189, 197
Westinghouse, George, 196, 198
  rivalry with Thomas Edison, 196
  Tesla and alternating current, 196–98
Westinghouse Company, 198
Whistler, George Washington, 78, 79
  Patrick locomotive engine, 78
Whistler, James A. M., 78
Wheeling Bridge, 209
Wilkinson, John, 22, 32
women
  first U.S. factory workers, at Lowell, 73

Yale College, 132, 136, 174